AGI ENVIRONMENT

COAL and the Environment

Stephen F. Greb

Cortland F. Eble

Douglas C. Peters

Alexander R. Papp

American Geological Institute

In cooperation with

Illinois Basin Consortium

U.S. Department of Energy, Office of Fossil Energy, National Energy Technology Laboratory

Office of Surface Mining

About the Authors

Stephen F. Greb is a research geologist with the Kentucky Geological Survey and adjunct faculty member of the Department of Earth and Environmental Sciences at the University of Kentucky. Dr. Greb has authored many papers on the depositional history of coal-bearing rocks, mining geology, and coal-related issues. He is a past-Chair of the Coal Division of the Geological Society of America and has won several awards for coal research. He is actively involved in outreach of geological information to the public. Recent research efforts have concentrated on geologic carbon sequestration.

Cortland F. Eble is a coal and energy geologist at the Kentucky Geological Survey; he also is an adjunct faculty member of the Department of Earth and Environmental Sciences at the University of Kentucky. Dr. Eble is a past-Chair of the Coal Division of the Geological Society of America and has won awards for coal research. He has numerous publications concerning palynology, the chemical and physical characteristics of coal, coal and its importance to the energy mix in the United States, and the fate of minerals and elements in coal from mining through utilization.

Douglas C. Peters is the owner of Peters Geosciences, a remote sensing and GIS consultancy in Golden, Colorado. He formerly was a Principal Investigator for the U.S. Bureau of Mines Denver Research Center, specializing in remote sensing and GIS applications for coal mining, abandoned mines, and environmental topic areas. Mr. Peters received M.Sc. degrees in Geology and Mining Engineering from the Colorado School of Mines. He is the author of more than 70 publications on coal geology, remote sensing, caving, mining, ground control, computer-aided geoscience, and GIS technology.

Alexander R. Papp has worked as a coal geologist for 25 years, both domestically and internationally, and held corporate, operations, and consulting firm positions. He has been involved in many phases of the "mining cycle" but principally in the collection of baseline environmental data, permitting activity, compliance assurance, and reclamation activities at exploration sites and mining operations. He received a M.Sc. degree from Eastern Kentucky University and is currently an independent consultant in Denver.

American Geological Institute
4220 King Street
Alexandria, VA 22302
(703) 379-2480
www.agiweb.org

The American Geological Institute (AGI) is a nonprofit federation of 44 scientific and professional associations that represent more than 120,000 geologists, geophysicists, and other earth scientists. Founded in 1948, AGI provides information services to geoscientists, serves as a voice of shared interests in our profession, plays a major role in strengthening geoscience education, and strives to increase public awareness of the vital role the geosciences play in mankind's use of resources and interaction with the environment. The Institute also provides a public-outreach web site, **www.earthscienceworld.org**.

To purchase additional copies of this book or receive an AGI publications catalog please contact AGI by mail or telephone, send an e-mail request to **pubs@agiweb.org**, or visit the online bookstore at **www.agiweb.org/pubs**.

Copyright 2006 by American Geological Institute
All rights reserved.

ISBN: 0-922152-77-2

Design: DeAtley Design
Project Management: Julia A. Jackson, GeoWorks
Printing: Ries Graphics

Contents

Foreword 4
Preface 5

1. It Helps To Know 7
Why Coal Is Important 7
What the Environmental Concerns Are 8
How Science Can Help 8
What Coal Is 8
Coal's Role in the Carbon Cycle 9
How Coal Forms 10
Resources and Reserves 11

2. Finding and Mining Coal 15
Exploration 15
Mining 16
    Underground Mining 17
    Surface Mining 17
Environmental Concerns 18
    Physical Disturbance 18
    Subsidence and Settlement 21
    Landslides 22
    Erosion, Runoff, and Flooding 23
    Water Quality 24
    Coal Mine Fires 28
    Fugitive Methane 29
    Safety and Disturbance Concerns 30
    Miners' Health and Safety 32

3. Transporting and Processing Coal 35
Transportation 35
Coal Preparation 36
Coal Processing 36
Environmental Impacts 37
    Road Damage and Public Safety 38
    Water Quality and Acidic Drainage 38
    Slurry Impoundments 39
    Physical Disturbances and Gob Fires 41

4. Using Coal 43
Power and Heat Generation 43
Impacts of Coal Use 44
    Sulfur Oxides and Acid Rain 44
    NO_x, Acid Rain, Smog, and Ozone 46
    Particulate Emissions and Haze 47
    Mercury and Hazardous Air Pollutants 48
    Carbon Dioxide 49
    Solid Waste Byproducts 51

5. Providing for the Future 53
Support for Technology Development 53
Future Electricity from Clean Coal Technologies 55
    Fluidized Bed Combustion 55
    Gasification Technology 56
    FutureGen 57
Liquid Fuels from Coal 58
The Future of Coal 58

References and Web Resources 60
Credits 62
Index 64

Foreword

Commercial coal mining began in the United States in the 1740s in Virginia, and coal fed our nation's industrial revolution and economic growth. Unfortunately, a long history of mining without regard to environmental consequences left a legacy of barren, disturbed landscapes and rust-colored, sediment-laden streams. Public realization of the environmental consequences of unregulated mining led to enactment of modern surface mining laws in 1977. Research into the environmental impacts of mining has resulted in a wide range of methods and technologies for cleaning up abandoned mine sites as well as for preventing and mitigating impacts from active mines. When mining is done properly, productive, long-term land uses are all that remains when the mining is completed.

Similarly, our increased use of coal for electric power led to unregulated emissions resulting in acid rain and increased haze in many parts of the country. Research into the types of emissions created from coal combustion, regulations, and new clean-coal technologies have reduced many harmful emissions; all while coal production has increased. More recently, an understanding that increased carbon dioxide emissions may contribute to climate change has resulted in a national initiative to create power plants with zero emissions that produce both power and useful byproducts.

Mining, processing, and using coal to meet our nation's energy needs while protecting natural environments will be an ongoing challenge. Many factors influence the potential impacts of coal extraction and use. Understanding the potential impacts and how they can be prevented, or mitigated, can help everyone meet this challenge.

This Environmental Awareness Series publication has been prepared to give educators, students, policy makers, and laypersons a better understanding of environmental concerns related to coal resources. AGI produces this Series in cooperation with its 44 Member Societies and others to provide a non-technical geoscience framework considering environmental questions. *Coal and the Environment* was prepared under the sponsorship of the AGI Environmental Geoscience Advisory Committee in cooperation with the geological surveys of Kentucky, Indiana, and Illinois (Illinois Basin Consortium), U.S. Office of Surface Mining, and the Department of Energy with additional support from the AGI Foundation and the U.S. Geological Survey. Series publications are listed on the inside back cover and are available from the American Geological Institute.

Travis L. Hudson, *AGI Director of Environmental Affairs*
Philip E. LaMoreaux, *Chair, AGI Environmental Geoscience Advisory Committee*

Preface

Coal, our most important domestic fuel resource, accounts for nearly 25% of our country's total primary energy production and produces half of our electric power. Annual U.S. coal production is 1.1 billion short tons, which equates to 20 pounds of coal per person, per day. On average you will use 3 to 4 tons of coal this year, probably without even knowing it.

That said, the U.S. Department of Energy indicates that because of the shear volume of energy our country needs to sustain economic growth and our standard of living, the use of coal as a fuel will likely increase in the future—even if the percentage of coal as a whole in the energy mix decreases. Increasing coal use is also expected in world markets as both China and India have large populations, rapidly expanding industrial economies and energy needs, and large coal resources of their own. The use of coal, like nearly all human activities, has environmental impacts. Recognizing these impacts has led to greater scrutiny in the way coal is mined, processed, and used.

Our objective in writing about coal is to relate the mining and use of this vital energy resource to the environmental concerns that affect our society. *Coal and the Environment* covers issues related to coal mining and combustion, as well as the methods, technology, and regulation currently in use, or planned for the future, to meet our nation's energy needs, while caring for the environment around us.

The authors gratefully acknowledge the many individuals who helped in putting this publication together. Special thanks to Travis Hudson and Julie Jackson for coordination and editing, and to Julie DeAtley for her phenomenal layout and design. Joe Galetovic, Office of Surface Mining, provided information and sources of images; Mark Carew and Ben Enzweiler, Kentucky Division of Abandoned Mine Lands, were also a great help in providing images. Thanks to all of the colleagues who provided technical expertise and images for use in the manuscript. We especially thank the principal reviewers for their time and efforts including James C. Cobb, Kentucky Geological Survey; Bob Finkelman, U.S. Geological Survey; Travis Hudson, American Geological Institute; Bob Kane, U.S. Department of Energy; Philip LaMoreaux, P.E. LaMoreaux & Associates; David Morse, Illinois State Geological Survey; Alma Paty, American Coal Foundation; John Rupp and Nelson Shaffer, Indiana Geological Survey; Gary Stiegel, U.S. Department of Energy; Steve Trammel, Kennecott Energy; and Dave Wunsch, New Hampshire Geological Survey.

Stephen F. Greb
Cortland F. Eble
Douglas C. Peters
Alexander R. Papp

June 2006

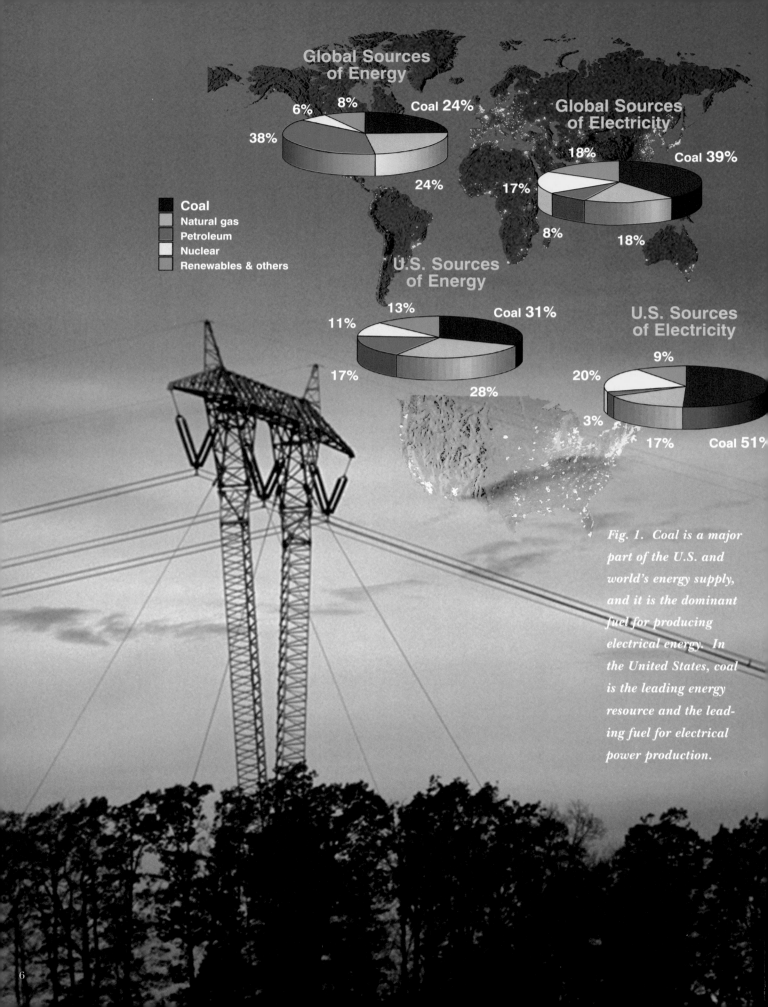

Fig. 1. Coal is a major part of the U.S. and world's energy supply, and it is the dominant fuel for producing electrical energy. In the United States, coal is the leading energy resource and the leading fuel for electrical power production.

It Helps to Know

Coal, the rock that burns, is an important source of global energy (Fig. 1). This fossil fuel formed from accumulations of plants under swampy conditions. The energy in coal originally came from the Sun, (through the plants) and when coal burns, energy is released.

Why Coal Is Important

Coal is the most abundant fossil fuel on earth, and U.S. reserves equal a quarter of the world's total (Fig. 2). The United States, former Soviet Union, China, Australia, and India have about 75% of the world's coal reserves. The global distribution of coal is different from that of petroleum — the Middle East has very little coal. At the present rate of consumption, Earth's coal reserves will last at least 200 years.

Although people have used coal to heat homes for hundreds of years, the major use of coal today is to generate electricity. In the United States, coal accounts for nearly one third of the country's total energy production and produces half of our electric power. Current annual U.S. coal production is 1.1 billion short tons, which equates to 20 pounds of coal per person, per day. On average you will use 3 to 4 tons of coal this year, probably without even knowing it. Providing this important source of energy involves different types of mining, processing, and technology; each associated with different environmental concerns.

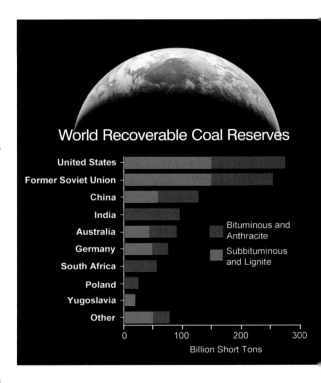

Fig. 2. The United States has 25% of the world's coal reserves. U.S. data from 2001; other countries from 2000.

Fig. 3

Coal, a natural hydrocarbon, consists primarily of rings of carbon (gray) and hydrogen (red) atoms bonded (lines) together. Other atoms such as sulfur (yellow) are trapped in coal along with gases and liquids.

What the Environmental Concerns Are

The long history of coal mining has left an unfortunate environmental legacy. Yet, that legacy helps us to understand the different ways mining, processing, and using coal can impact the environment so that future impacts can be mitigated or prevented, including

- Disturbances of the landscape;
- Water quality;
- Air quality;
- Combustion waste management; and
- Public health and mine safety.

How Science Can Help

A sound understanding of the physical and chemical processes that take place during the mining, processing, and use of coal is helping to identify, minimize and/or mitigate undesirable environmental impacts. Coal mining and processing operations in the United States use a wide array of methods to limit environmental impacts including

- Scientific studies to identify potential environmental impacts before mining and processing begin;
- Better engineering and scientific designs that help prevent or minimize impacts off site; and
- Modern reclamation techniques for returning disturbed mine lands to environmentally acceptable uses.

Coal-fired electric utilities also face environmental challenges that science and technology address through

- Applying technologies that increase fuel efficiency while decreasing potentially harmful emissions;
- Conducting research and development of technologies that help to reduce emissions of sulfur dioxide (SO_2), nitrogen oxides (NO_X), and particulate matter from power plants;
- Continued scientific research and development of new technologies to capture and permanently store emissions, such as carbon dioxide, that cannot be reduced by other means; and
- Developing uses for coal combustion wastes to reduce the amount placed in landfills or impoundments.

What Coal Is

Coal contains abundant amounts of carbon, a naturally occurring element that is common in all living things. Combined with hydrogen, the two elements form a group of widely varied compounds called hydrocarbons (Fig. 3). Although coal contains some gases and liquids it is a solid hydrocarbon. Natural hydrocarbons, including coal, are called "fossil fuels" because these fuel sources originated from the accumulation, transformation, and preservation of ancient "fossil" plants and, in some cases, other organisms.

Most coal is formed from the remains of plants that accumulated under swampy conditions as peat (Fig. 4). Imprints of fossil stems, roots, and leaves are common in coal and surrounding sedimentary rocks.

Fig. 4. The painting depicts a likely setting for coal formation 300 million years ago. Coal forms from peat that accumulates under wetland conditions, but not all swamps or wetlands will lead to coal formation.

Plant fossil

Ancient coal swamp

Modern peat bog, *Alaska*

WHERE COAL FORMS

However, it takes a great amount of carbon-rich plant material, time for that material to form peat, and special geological and chemical conditions that protect the peat from degradation and erosion to make a mineable coal seam. Peat and the buried coal that eventually forms from it are part of our planet's carbon cycle.

Coal's Role in the Carbon Cycle

Carbon is cycled through the earth in several forms — for example, as part of the atmosphere, or in living organisms as part of the biosphere (Fig. 5). Plants absorb carbon dioxide (CO_2) from the atmosphere during photosynthesis,

Modern swamp, *Florida*

Carbon Cycle

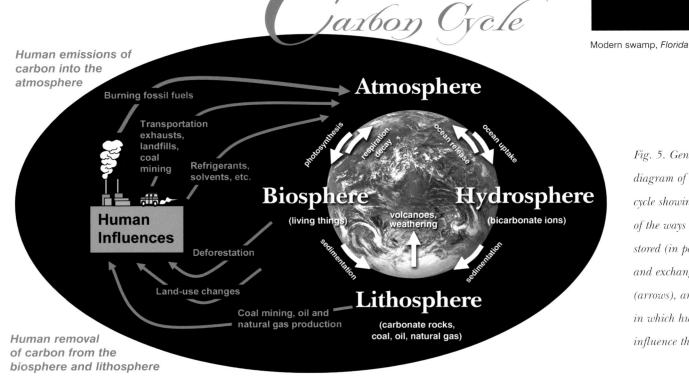

Fig. 5. Generalized diagram of the carbon cycle showing some of the ways carbon is stored (in parentheses) and exchanged (arrows), and the ways in which humans influence the cycle.

Fig. 6 Carbon dioxide is considered a "greenhouse gas" and increased levels of CO_2 and other greenhouse gases in the atmosphere may contribute to global warming. According to the U.S. Environmental Protection Agency's inventory of greenhouse gas emissions (2004), the major greenhouse gases put into the air by human acitvities (in carbon dioxide equivalents) are

carbon dioxide (CO_2)   85%
methane (CH_4)   8%
nitrous oxide (N_2O)   5%
and other gases   2%

and release CO_2 back into the atmosphere during respiration. Carbon from photosynthesis is stored in the plants. If the plants die and accumulate as peat, the precursor of coal, the carbon becomes part of the geosphere. Peat that is buried and transformed into coal is a vast carbon sink or reservoir.

Coal deposits store carbon in the geosphere for millions of years and are long-term carbon sinks. When coal and other long-term carbon sinks are removed from the geosphere through mining or other human activities we disrupt the natural carbon cycle. Burning coal or other fossil fuels oxidizes carbon, produces heat, and releases byproduct carbon dioxide (CO_2) into the atmosphere at a rate faster than would occur naturally. Greenhouse gases, such as carbon dioxide and methane, act as an insulating blanket around the Earth, allowing incoming solar radiation to warm the Earth's surface and reducing radiation of heat back into space (Fig. 6). Because CO_2 is a greenhouse gas, there is concern that man-made increases in carbon emissions are rising, and contributing to global climate change. The role of coal combustion's possible influence on global climate is discussed in Chapter 4.

How Coal Forms

Large amounts of plant materials accumulate in widespread peat-forming wetlands (called mires). When mires accumulate within geologic basins, they can be deeply buried long enough for the peat to be converted to coal (Fig. 7). Basins are broad, subsiding (sinking) depressions in the Earth's crust in which sediments accumulate.

When peat is buried, pressure from the overlying sediments and heat

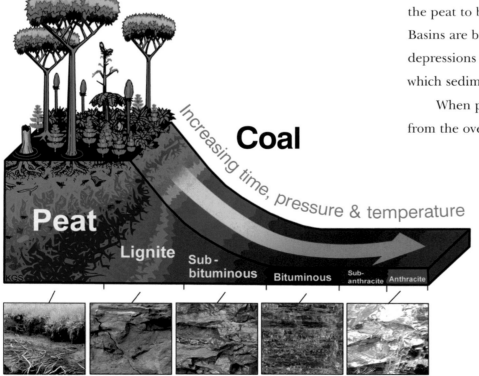

Fig. 7. Coal "rank" (the stage of coal formation) increases from peat to anthracite with time, heat, and pressure.

within the Earth transforms the peat chemically and physically into coal. This process, called "coalification," results in several types or stages of coal. These stages of coal formation are classified as "rank". The ranks of coal, in increasing alteration from peat, are lignite (brown coal), sub-bituminous, bituminous, semi-anthracite, and anthracite. If coal is heated beyond the rank of anthracite, it becomes a form of almost pure carbon (graphite or natural coke). Higher rank coals produce more heat per ton when they burn than lower rank coals because they are more concentrated forms of carbon. Put another way, one must burn more low-rank than high-rank coal to produce the same amount of energy.

During coalification, compaction and dewatering cause fractures to form in the coal. These fractures are called "cleats." Water moving through porous peat or through cleats in coal can carry and deposit minerals. Some of the most common minerals in coal are silicates (quartz, clays), carbonates (calcite, siderite) and sulfides (pyrite, marcasite). The elements within these minerals (for example, sulfur) may cause environmental concerns during the processing and burning of coal.

Resources and Reserves

Coal is mined throughout the United States (Fig. 8). The Powder River Basin in Wyoming and Montana, the Central and Northern Appalachian basins, and the Illinois Basin (also called the Eastern Interior Basin) are the largest coal producing regions. Differences in geology, geography, and climate between basins mean that the mining and use of coals from each of the several basins have unique environmental concerns. Coals from some areas must be processed before they are used; other coals can be used without processing other than handling and loading for transport.

Western coals are lower in sulfur content than Interior and some Eastern coals, and the western deposits are thick, near the surface, and easily accessible. For these reasons, coal production from large open pit mines in the West has increased coal production. Wyoming is currently the nation's leading coal producer, accounting for a third of U.S. coal production.

The top three producing states, Wyoming, West Virginia, and Kentucky, account for more than half of the country's annual production, but 20 states each have demonstrated reserves of more than 1 billion tons. Demonstrated reserves are estimates of the amount of coal in the ground that has been measured with a relatively high degree of confidence and which is technically recoverable under current economic conditions. The definition of coal resources is broader and includes the total estimated amount of coal in the ground. Resources consist of demonstrated reserves plus coals that might not be currently mineable or for which there is less data and therefore lower confidence in their thickness or distribution.

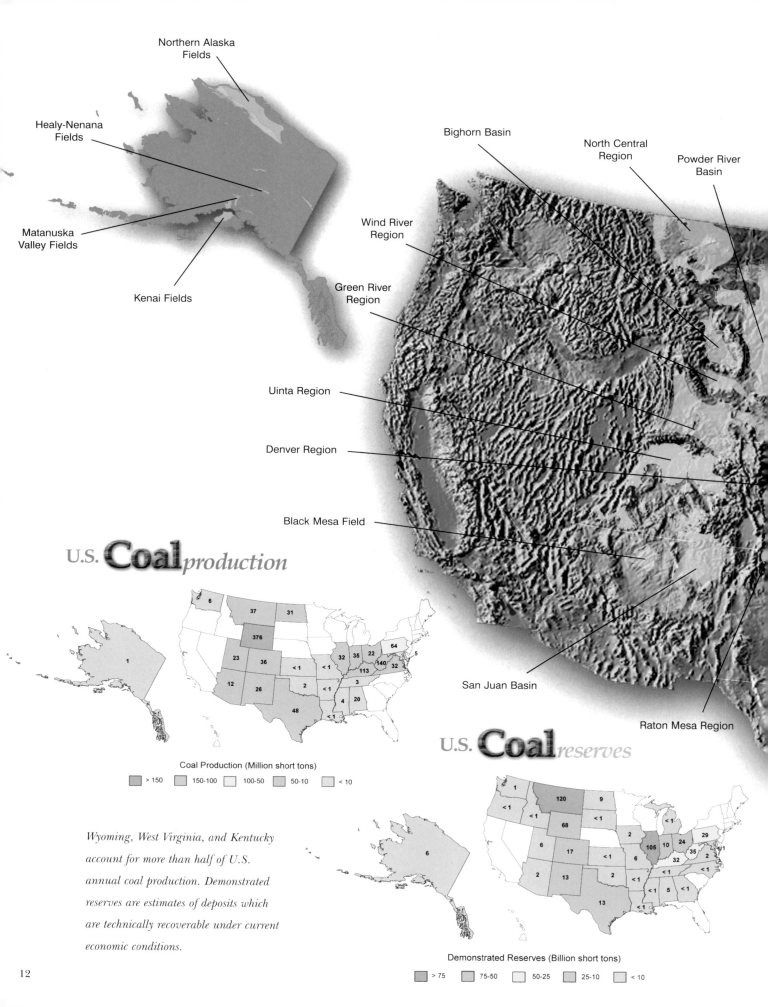

U.S. Coal *production*

Coal Production (Million short tons)
> 150 | 150-100 | 100-50 | 50-10 | < 10

U.S. Coal *reserves*

Demonstrated Reserves (Billion short tons)
> 75 | 75-50 | 50-25 | 25-10 | < 10

Wyoming, West Virginia, and Kentucky account for more than half of U.S. annual coal production. Demonstrated reserves are estimates of deposits which are technically recoverable under current economic conditions.

Fig. 8. Mining and use of coals from each basin presents unique environmental concerns, due to differences in geology, geography, and climate. The Powder River Basin, Central and Northern Appalachian basins, and the Illinois basin are the largest coal producers.

EXPLORATION   PLANNING   EXCAVATION   RECLAMATION

Fig. 9. The mining cycle starts with exploration and continues to closure and reclamation. When coal is mined at the surface, it is generally blasted and then extracted by shovel like this example from a Wyoming surface mine. When coal is mined underground, mechanized cutting machines extract it, as in this longwall mine in Colorado.

From PHYSICAL DISTURBANCE through POST MINE LAND USE

Physical disturbances to the landscape occur during surface mining and remain until the mined area is reclaimed. Since 1977, strict regulations have guided the reclamation process.

Before

After

(Before) Landslides and flooding from waste piles at this pre-1977 abandoned mine site in Virginia threatened homes down slope. (After) Reclamation included grading the piles, constructing drainage channels, adding topsoil, and establishing a vegetative cover to stabilize the slope.

Regrading disturbed land

Revegetating mined lands

Returning land to approved post-mined land uses

For example, roads are built, electrical and phone lines are brought to the site, offices and maintenance facilities are constructed. Other disturbances are specific to mining. For underground mines, the area of direct physical disturbance is generally small and concentrated around the entrance to the mine. For some underground mines, elevators and conveyor belts are built to transport miners and coal. Conveyor belts may extend far underground and above ground from the immediate mine entrance.

In contrast, surface mines have a broader footprint during mining because vegetation is removed prior to mining, and large amounts of rock must be removed to get to the coal. The amount of material removed depends on the type and scale of mining. Spoil material remains visible at the surface and is disposed of in accordance with the mine permit, usually as fill during reclamation. As surface mining progresses through an area, parts of a surface mine will be undergoing active mining, while other parts are being reclaimed so there is generally activity across a large area.

Physical impacts remain on the landscape until the mined area is reclaimed. Good reclamation plans restore the disturbed surface area for post-mine land uses and control runoff to protect water quality. Reclamation plans are required before any mining takes place and the plans must meet state and federal regulations. These plans must also include a post-mine land use agreed upon by the mine operator and landowner.

The type of reclamation undertaken at coal mines depends on the permitted post-mine land use, type of mining, the size or area of the disturbance, topography, and climate of the mine site. Reclamation bonds are posted to insure fulfillment of reclamation plans. These bonds are money (insurance policies) that mines must set aside with the appropriate regulatory agencies prior to mining so that if something happens to the mining company, money will be available to complete reclamation. The bond is generally released in phases, which are defined by regulators. Bonds are not fully released until the regulatory authorities are satisfied that all surface disturbances at the mine site are reclaimed.

An important initial step in reclamation is preservation of topsoil. When mined areas are first excavated, topsoil is segregated and banked in storage areas, and when mining is finished, the topsoil is replaced to facilitate revegetation. In most cases, the surface of the mined lands must be graded to nearly its original shape (termed approximate original contour). In the arid west, complex slopes that were not part of the original landscape are sometimes permitted in reclamation to limit wind erosion.

Where mountaintop mining methods are used, the land cannot be returned to approximate original contour, but reclamation does include grading and revegetation. Debates about mountaintop removal include concerns about landscape changes, the extent of those changes (many square miles), the potential for increased sedimentation, and potential for surface water quality changes in the streams that drain the mine. During mountaintop removal, valleys are filled with the rock that is excavated to get at the coal. Several law suits have claimed that valley filling during mountaintop removal violates sections protecting streams in the 1972 Clean Water Act and the 1977 Surface Mining Control and Reclamation Act. These sections prohibit disturbing land within 100 feet of intermittent or perennial streams unless a variance is granted. Legal issues involving mountaintop mining continue.

An important part of the reclamation process is revegetation (Fig. 15). The types of vegetation permitted depend on site conditions, such as climate, elevation, and slope; the type of vegetation present before mining; wildlife; soil properties; and the permitted post-mine land use. Establishing good vegetative cover aids in controlling erosion and sedimentation (siltation), reducing water movement to the underlying mine spoil, decreasing oxygen concentrations, and increasing the capacity for carbonate dissolution, which can also aid in reducing or preventing acidic drainage.

Mining companies are required to establish a successful vegetative cover before their bonds are released; the time period is defined as a minimum of 5 years in the East and a minimum of 10 years in the West. Some companies are choosing reforestation as a post-mine land use, because it adds ecological benefits, such as limiting erosion and providing wildlife habitats. Planting forests also provides future, renewable timber resources and offers the added attraction of removing carbon dioxide from the atmosphere at a time when there is significant concern about rising CO_2 levels.

Fig. 15. Care is taken in choosing species tolerant of climate conditions in reclamation, like these native Kayenta pinon pines in Arizona.

Subsidence and Settlement

Sinking of the land surface caused by settlement of mine spoil in some mined areas, or by the collapse of bedrock above underground mines is called subsidence. Subsidence above underground mines occurs when the rock above mines collapse, resulting in bending and breakage of overlying strata that ultimately reaches the surface (Fig. 16). Settlement above mine spoil generally occurs because of compaction or dewatering of mine fill material through time.

Whether or not there will be subsidence impacts at the surface depends on the geology of the bedrock, depth of mining, and manner in which the coal

Fig. 16. At this church in western Kentucky, steel beams and wood cribs were erected to prevent further subsidence above an underground mine.

Fig. 17 LANDSLIDES

A retaining wall is built to protect this home from abandoned (pre-1977) mine slopes above.

During modern reclamation, material is pushed up against the mine highwall to approximate original contour of the landscape, the surface is vegetated, and water is redirected to prevent instability.

was extracted. Where room- and pillar-extraction is used, large blocks or "pillars" of coal are left between "rooms" where the coal was removed. If subsidence occurs, it will be localized above rooms and may occur at any time after mining. In contrast, longwall extraction methods remove coal in long panels and the overlying roof rock is designed to collapse in safe, controlled collapses behind the advancing panels. With longwall extraction, subsidence effects are more immediate and predictable. Subsidence can also occur as a result of underground mine fires regardless of mining methods or in the absence of mining (see p. 28).

It is estimated that nearly 2 million acres (8,000 km^2) of land have been affected by subsidence above abandoned (pre-1977) coal mines in the United States. Recognition of past subsidence problems led to federal and state guidelines that restrict underground mining, and generally limit or prohibit mining beneath towns, major roads, and waterways.

Since 1977, more than 2,000 subsidence problems have been corrected through the Abandoned Mine Land Emergency Program. Stabilization is generally achieved by drilling into the abandoned mines and pumping cement or concrete-like materials into the mine voids.

Landslides

Landslides are a concern in coal mining areas with steep topography. Regions of steep topography are prone to natural landslides and slope failure, but mining can increase the likelihood of slope failures by removing vegetation from the hillside; disrupting the base (toes) of natural, pre-existing slumps during mining and road construction, and redirecting surface and groundwater in ways that saturate naturally unstable slopes (Fig. 17). The Surface Mine Control and Reclamation Act (1977) set standards for surface mining that include returning mined areas to near their natural slope (termed approximate original contour) to avoid landslides and slope failures.

Prior to this legislation, more than 8,600 acres of dangerous slides were identified at abandoned coal mines. Since 1977, the U.S. Department of the Interior's Office of Surface Mining and associated state regulatory agencies have reclaimed 800 known slope failures on more than 3,400 acres of mined lands. Mitigation of mine-induced slope failures generally involves redirecting water away from slump-prone areas. Disturbed areas are then graded and revegetated. In some cases, retaining walls are built to protect structures, such as roads and houses, which are located downslope from known landslides.

Reclamation of highwalls in active or abandoned surface mines involves backfilling rock against the highwall, and compacting and grading the fill material to minimize future slumping and sliding. Backfilled slopes are then revegetated to prevent slope failures.

Erosion, Runoff, and Flooding

Changes in drainage and sedimentation are common environmental concerns in any excavation or construction site including surface mines. Increased sedimentation can degrade water quality, smother fauna at the bottom of streams and lakes, fill lakes and ponds, act as a carrier of other pollutants, and clog stream courses, which can lead to flooding. In the past, substantial increases in sedimentation resulted from deforestation of mine areas prior to mining.

In modern mining, sedimentation is controlled through better forest harvesting practices prior to mining, ongoing reclamation that limits the amount of disturbed material at any one time, construction of roads with culverts and buffers to limit or direct runoff, and the use of terraces and grading to reduce steep slopes, which limits erosion and controls or directs runoff. Sediment ponds are required at all mine sites to trap sediment and prevent it from leaving the site (Fig. 18). Once the sediment settles out, the water can be discharged into downstream waterways. During mining, settling ponds are routinely dredged and the dredged material is added to the mine spoil.

Sedimentation concerns are different in arid western states. Thin vegetative cover, flash floods, and wind erosion make

Fig. 18. Sediment ponds are constructed at surface mines to trap sediment-laden waters and prevent sediment from leaving the mine site. The rock drain in the upper photo directs the flow of mine waters to sediment ponds at a mine in Indiana. The pond and wetland in the lower photo were created during reclamation of a surface mine in Texas to provide flood storage.

Fig. 19. The water quality of surface streams on mine sites is analyzed before, during, and after mining. The scientists in the photo are counting fish in a stream on abandoned mine land as a measure of the stream's health. The sample shown is being tested to determine its pH, the degree of acidity or alkalinity.

arid landscapes especially susceptible to erosion. In such areas, the goal of inhibiting erosion must be coupled with retaining available moisture if sedimentation is going to be limited and revegetation successful. Some of the practices used to prevent erosion and sedimentation from western mining include digging furrows, constructing check dams, contour terracing, lining drainage channels with rock and vegetation, and mulching.

Water Quality

Mining results in large increases in the amount of rock surfaces exposed to the air and water. In spoil piles or backfill, the newly exposed rock surfaces reacting to air and water may lead to changes in the

- Acidity, pH;
- Sediment load;
- Total suspended solids; and
- Salinity (total dissolved solids)

of the water passing through the disturbed material. In order to track potential water quality changes resulting from mining, coal companies must monitor all surface and groundwater on their sites before, during, and after mining. Water standards are set by federal, state, and tribal authorities. Some of the parameters tested to determine if mining is altering off-site water quality include pH, conductivity, dissolved oxygen, total suspended solids, total dissolved solids, including bicarbonate, nitrate-nitrite, phosphate, and varied elemental (iron, manganese, etc.) concentrations (Fig. 19).

Mine-related, surface-water quality issues depend in part on climate. In the arid western states, production of alkaline (high pH) waters with increased total dissolved solids is a potential consequence of disturbing surface materials naturally rich in sodium and calcium sulfates. Likewise, leaching of trace elements that are soluble in alkaline waters, such as boron and selenium, is a concern, because high concentrations of these elements can be toxic to plants and animals. To prevent these consequences, regulatory agencies developed a series of best practices to limit the production and downstream migration of alkaline waters from western coal mines. Some of these practices include

- Computer modeling to better implement site-specific sedimentation and erosion plans and technology;
- Use of terraces, contour berms, diversion channels, and check dams to control runoff and erosion;
- Regrading and complex slope design to limit erosion and runoff;
- Mulching to increase infiltration and retain water; and
- Roughening, pitting and, contour plowing, to increase infiltration and aid in revegetation.

Acidic Drainage

Acidic (low pH) waters are a particular concern in the eastern United States, where a longer unregulated mining history, climate, and rock characteristics plus the population density around impacted waters make acidic drainage a major environmental issue. Water from mined lands with increased acidity, and higher concentrations of dissolved metals, especially iron, aluminum, and manganese (Fig. 20) can be a problem.

Fig. 20. The orange-colored water leaking from an abandoned mine opening is characteristic of acidic (acid rock) drainage. Pyrite, oxygen, and bacteria are the main ingredients that combine in nature to make the sulfuric acid that acidifies soil and water. Acidic drainage results from mines located in areas that contain strata and coal with high pyrite and low carbonate concentrations.

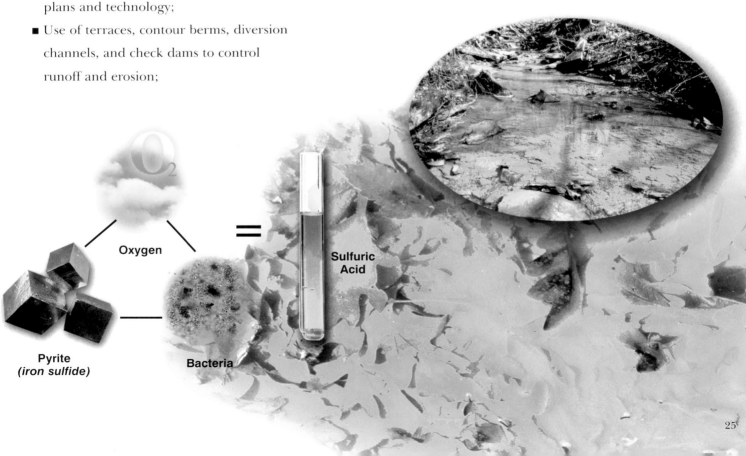

Acidic drainage does not result from every mining operation, but rather, from mines located in strata and coal with high pyrite and low carbonate concentrations. Acidic drainage also occurs from un-mined exposures as a natural consequence of weathering. Pyrite, commonly called "fool's gold," is an iron-sulfide mineral, which may be present in high concentrations in coal beds and organic-rich shale. The reaction of pyrite with oxygen in soil, air, or water is the principal cause of acidic drainage.

Acidic drainage can result in depleted oxygen levels, toxicity, corrosion and precipitates that can degrade water quality, damage aquatic habitats, and can make surface and groundwater unusable for post-mine land uses.

Modern surface mining techniques have greatly reduced the amount of acidic drainage produced by mining. If neutralizing materials (such as limestone) occur within the material that will be mined, they are mixed with potentially acidic rock strata to neutralize acidic water produced. Rock layers identified as containing high percentages of pyrite are removed selectively and disposed of in a manner that limits further oxidation or surface runoff. Selective handling is combined with

- spoil placement above the water table;
- diversion of waters away from the material;
- treatment to reduce acidity in runoff (where needed); and
- covering with sealants (such as clays)

in order to prevent interaction with groundwater and surface water.

Although modern mining companies spend great effort preventing acidic drainage, there is an unfortunate legacy of acid-rich, rust-colored, and biologically impaired streams resulting from past mining. The U.S. Environmental Protection Agency estimates that acidic drainage has polluted 17,000 km (10,874 miles) of streams in Appalachia. Many methods have been developed to mitigate this legacy. No single method is appropriate for all situations (Fig. 21). The most common method for treating mine-caused acidic waters are so called "active" techniques in which neutralizing material, such as limestone, is continuously added to affected waterways through a water treatment facility or similar procedure. Engineered structural techniques are also common and include various methods of water management to redirect or divert water from potentially acid-producing material. Other remediation methods include "passive" treatments that do not require chemical additions, but use natural chemical and biological processes to reduce acidic drainage. Examples of passive treatments include

- Constructed wetlands;
- Anoxic limestone drains; and
- Successive alkalinity producing systems.

Fig. 21 ACIDIC DRAINAGE REMEDIATION

Limestone is alkaline and is a common "active treatment" used to neutralize acidic drainage.

Analyzing drainage

Directing drainage

Constructed wetland

Limestone drains

Vertical flow ponds

Groundwater Protection

A principal environmental concern associated with mining any material from beneath the surface, including coal, is groundwater or aquifer protection. Groundwater is water that moves through rock layers beneath the surface of the earth. Groundwater-bearing rock layers that can produce enough water to be used as a water supply are called aquifers. Mining can impact groundwater in several ways. Water passing through soils, mined areas, and spoil can pick up soluble elements (mostly salts including sulfates, calcium, and magnesium) to form leachates. These solutions can leak through fractures and enter shallow groundwater aquifers, causing increases in total dissolved solids. Likewise, surface mines and abandoned underground mines can be the source of acidic drainage, which can move into aquifers through fractures.

To determine if mining is influencing groundwater in mining areas, monitoring wells are emplaced in known aquifers and sampled at intervals determined by regulatory authorities (Fig. 22). Some of the parameters tested during groundwater monitoring include temperature, pH, specific conductance, acidity, alkalinity, total dissolved solids, carbonate, bicarbonate, trace elements, nitrogen species, and total suspended solids.

Fig. 22. Wells are drilled around a mine site so that groundwater can be periodically checked for any negative impacts that might be caused by mining.

Fig. 23

COAL MINE FIRES

Citizens of Centralia, Pennsylvania, were relocated because of hazards from an underground mine fire that is still burning.

Mining can also impact the amount of available groundwater. In many interior and western states, surface-mined coals (which are also the shallow aquifers) and sediment or rock above the coal must be dewatered to allow mining. Dewatering means that water is pumped out of the coal and surrounding rock. The pumping can lead to a short-term decrease in water levels in shallow wells near the mine. Dewatering and water use by mining have caused local concerns in some western states because of increasing competition for limited water supplies.

In areas of subsidence above abandoned underground mines (see page 21) groundwater flow and storage capacity can be changed, leading to decreases in local yields of water wells or changes in water chemistry.

In most states, if mining leads to changes in water levels or quality in wells adjacent to the property, the mining company must install new wells into a deeper, unaffected aquifer, or provide another source of water.

Coal Mine Fires

Underground coal fires have been among the worst disasters in U.S. coal mining history. Coal fires are started by various means including lightning, forest fires, spontaneous combustion, accidental fires started during mining, and ignition (man-made or natural) of mine refuse and other materials adjacent to outcrops of coal. Fires in coal beds burn slowly (tens of meters per year), but they can burn for decades. Coal fires can cause unsafe heat, forest fires, noxious emissions, and surface subsidence (Fig. 23). Subsidence can occur when the coal and surrounding rocks are baked by the fire, which causes the strata to compress or compact, and results in collapse of the overlying material.

It is difficult to determine the extent of underground coal fires, and such fires are very difficult to extinguish. To extinguish an underground mine fire you have to eliminate the fuel (the coal), heat, or oxygen. Several fire control techniques are used and the determination of which technique is used depends on the risk to adjacent property, original mining type, local geology and hydrology. Eliminating the fuel requires complete excavation of the coal or digging a trench or constructing a barrier to prevent the spread of the fire. Eliminating the heat usually involves flooding or flushing the fire area with water. Eliminating the flow of air and oxygen to the fire generally requires flushing mine voids with a slurry of water and fine particles to plug pore spaces, cleats, and fractures, and surface sealing of abandoned mine openings to eliminate ventilation of the fire farther underground.

Fugitive Methane

Fugitive methane is the uncontrolled release of methane to the atmosphere. Methane (CH_4) is a naturally occurring gas in coal that forms from anaerobic methanogenic bacteria and chemical reactions of coalification. The amount of methane in a coal depends on the coal's rank, composition, age, burial depth, and other factors. When coal is mined, the gas trapped within it is released.

Methane has long been a concern in terms of miner safety. Some of the worst U.S. mining disasters are caused by methane explosions in underground mines. Fugitive methane can also be a hazard at the surface if it leaks from underground mines (active or abandoned) through fractures into buildings and water wells.

In order to prevent explosions of methane (or methane and coal dust combined) methane concentrations are constantly monitored and large exhaust fans are used to circulate fresh air from the surface into the mine. Methane becomes part of the exhaust air and is generally vented to the atmosphere. Coal that is left exposed underground (for example, pillars in room-and-pillar mines) is covered with powdered limestone (called "rock dust") or other non-combustible material to keep a blast from spreading, and to keep coal dust from becoming suspended in the mine air (coal dust in the air is explosive). If methane leaks to the surface during or after mining, remediation generally focuses on mitigation at the point of concern by redirecting, venting, or sealing the path of the escaping gas.

Because methane is a greenhouse gas, there is also concern that anthropogenic emissions of methane may contribute to global climate change.

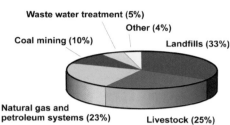

Fig. 24. This geologist is sealing a coal core that has just been drilled in a canister for measuring the coal's methane gas content. The chart shows human-related sources of methane in the United States.

Landfills and agriculture account for most of the anthropogenic methane released in the United States; coal mining accounts for 10% (Fig. 24). Shifts in U.S. production to western surface-mined coals and recovery of methane as a fuel have led to decreases in fugitive methane from mining of more than 30% since 1990. Unlike other greenhouse gases, methane can be used as a clean, hydrogen-rich fuel source. Therefore the principal method for mitigating methane releases from underground coal mines is to drill into the coal in advance of mining and collect the methane. New capture technologies to harness ventilation methane are being researched and developed. Use of these technologies is not practical or economic in all coal basins. In some cases, coal-bed methane is a primary energy resource, produced from coal beds that cannot be mined. In fact, one method being investigated to decrease the amount of anthropogenic carbon dioxide released is to store it in deep, unmineable coal beds (also called sequestration, see p. 50) and use the carbon dioxide to drive out the coal-bed methane for use as fuel.

Safety and Disturbance Concerns

Several of the environmental issues related to coal mining are also related to public disturbance, welfare, and safety. Blasting and dust are probably the most common nuisance or disturbance issues. Surface mines use explosives to break rock layers above the coal, and sometimes the coal itself (Fig. 25). Blasting is a safety issue because fatalities, injuries, and property damage have occurred from coal-mine blasting accidents. Blasting and vehicle movement at mines also produces dust. Dust can limit visibility and is a health concern because long-term (chronic) exposure to high levels of mine dust can cause respiratory problems.

Regulations set limits on dust and vibration levels in modern mines. To limit dust, mines spray water (from special water tank trucks) on all active road surfaces. Mining companies also revegetate disturbed areas and exposed spoil piles to prevent dust formation. To limit damage from blasting, all dwellings within a half-mile of proposed mine sites are identified prior to mining and appropriate blasting levels are calculated to prevent damage to dwellings. Notices of blasting schedules, signs and warning sirens are required during blasting and all blasting must be done by state-certified blasters. Noise levels and vibrations are monitored by the mining companies and must meet State and Federal regulations. If mine blasts cause damage to property, the mining company

Fig. 25

PUBLIC SAFETY HAZARDS

Blasting and dust from active mines

Pre-1977 abandoned buildings and equipment

Blowouts from abandoned mines

Dangerous highwalls at pre-1977 abandoned mines

must repair the damage or otherwise settle with the property owner.

An array of potential dangers are associated with abandoned mines. Some of the features that can pose dangers are abandoned highwalls, impoundments and water bodies, open portals (mine openings) and shafts, hazardous equipment and facilities, and illegal dumps. Old mine openings are usually sealed or barricaded, but sealed mine openings are sometimes reopened by those seeking adventure or those looking for a local coal supply. Such adventures are inherently dangerous, as abandoned mines are no longer ventilated and therefore may have low-oxygen areas, poisonous or explosive gas concentrations, flooded sections, and areas of unstable roof.

Abandoned mine sites are also potentially dangerous, especially to the curious, or adventurous because of impounded water in abandoned surface pits and old rusted mining equipment and building structures. Likewise, water can accumulate within abandoned underground mines. The size of the mine (and open space), slope of the mine, and amount of water entering the mine determine how much water can accumulate. If large abandoned mines are above drainage (above the lowest level of streams in an area), there is a potential hazard from blowouts (breakouts). Blowouts occur when the water pressure in flooded mines exceeds the strength of the seals placed at old mine openings or barrier pillars. Such blowouts were once common in Appalachian coalfields, resulting in catastrophic flooding downstream. State laws have resulted in better seals and barriers that significantly reduced the number of blowouts, but they still occur. In April of 2005, a blowout in eastern Kentucky flooded and damaged part of a major state highway, causing the highway to be shut down for several days, until water levels from the mine decreased (Fig. 25).

Miners' Health and Safety

Mining is a difficult and potentially dangerous profession. In a single year, 1907, 3,242 coal miners were killed in U.S. coal mines. Increasing use of technology, improved mining methods, increased miner education and training, and regulatory oversight has dramatically improved the safety of U.S. coal mines. In 2005, 22 fatalities were reported (Fig. 26). There is still much progress to be made in reducing fatalities, injuries, and illnesses in coal mines but the progress U.S. mines have made in safety stands in dramatic contrast to some developing nations, in which thousands of miners are still killed annually in coal mines.

Although black lung and silicosis are declining in the United States, these diseases still impact coal miners. Black lung disease is a hardening of the lungs caused from prolonged inhalation of coal dust. The disease mostly affects miners over the age of 50 who have had long-term exposure to excessive mine dust. Silicosis is a lung disease resulting from the long-term inhalation of silica dust from rock drilling or during loading and transport of rock materials.

In order to decrease the occurrence of both of these dust-related diseases, stringent regulations on the amount of inhalable dust are placed on underground mining operations. All dust control plans are approved by the Mine Safety and Health Administration before mining begins, and dust levels must be monitored and verified throughout mining.

Fig. 26

MINE SAFETY

Robots are being developed to travel through unsafe parts of mines when needed.

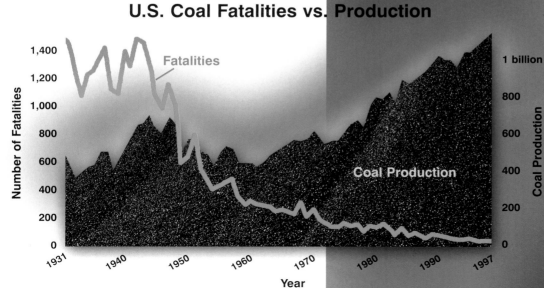

U.S. Coal Fatalities vs. Production

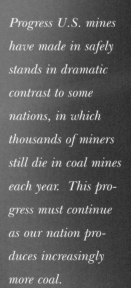

Progress U.S. mines have made in safely stands in dramatic contrast to some nations, in which thousands of miners still die in coal mines each year. This progress must continue as our nation produces increasingly more coal.

Fig. 27. Coal may be moved by rail, barge, conveyor, and truck. Sixty to 70% of the coal mined in the United States annually is shipped by rail.

3 Transporting & Processing Coal

Once coal is mined, it must be transported to the user. In many cases, the coal is processed to remove impurities prior to shipping. Understanding the way coal is transported and processed is important to anticipating, mitigating, or preventing potential adverse environmental impacts.

Transportation

Coal is primarily shipped by rail, truck, barge, and ship (Fig. 27). In underground mines, conveyor belts are used to transport coal to the surface, and in some cases, power plants are near-enough to mines that conveyor belts bring the coal directly to the plant. In most cases, however, multiple modes of transportation are used. Coal is commonly transported from different mines by conveyor or truck to a centralized loading facility where the coal is shipped by rail or barge. Sixty to 70% of the coal mined in the United States is transported by rail. Rail is especially important for shipping western U.S. coal to power plants in the Midwest and East. States regulate river and rail transport, including loading facilities. In the eastern United States, much of the coal mined is transported by trucks for at least part of its journey to the power plant. Roads used by trucks hauling coal are termed "coal haul roads" and are designated as acceptable transportation routes by individual states. Weight limits for trucks using those roads are also set by individual states. For example, in Kentucky the legal weight limit is 120,000 pounds.

Fig. 28. In preparation plants, coal is processed in multiple stages to remove impurities, such as the pyrite shown in the microscopic image of a high-sulfur coal. Slurries of magnetite are commonly used to adjust the density of liquid to remove impurities, like pyrite, in the coal. The magnetite is recovered as part of the coal cleaning process.

In the arid west, competing uses for water are a local public concern where water is used to transport coal as slurry. Some coal in the southwest is mixed with water and piped as slurry from the mine or preparation facilities to power plants. The Black Mesa pipeline in northeastern Arizona is the longest coal pipeline in the United States; each year 4.8 million tons of coal travel through the 273-mile long pipeline en route to a power plant in Nevada. Because this pipeline passes through Native American reservations and it crosses state lines, it is regulated by state and federal statutes.

Coal Preparation

During the mining process, some rock and clay from above and below the coal seam may be recovered along with the coal. The coal itself also contains rock and mineral impurities (ash) that will not burn. This "run-of-mine" coal may be processed or "cleaned" to control particle size, increase the relative heating value of the coal and remove the mineral components from the organic combustible components. This processing is referred to as coal preparation, beneficiation, cleaning, or washing, and is accomplished in special facilities called preparation plants. Approximately half of the U.S. bituminous coal mined annually is processed.

Coal Processing

Coal processing has become increasingly important in coal fields with medium- and high-sulfur coals since the enactment of the Clean Air Act Amendments (1990); the amendments required markedly lower sulfur dioxide emissions from power plants.

Several levels of coal processing are available, depending on the quality of the run-of-mine coal and needs of the end user. Initially, raw coal is crushed to separate large particles of unwanted mineral matter from combustible organic matter. Next, crushed coal is sized into coarse and fine coal fractions. Various types of screens and sieves may be used to size particles prior to cleaning. Coarse material is put through density (heavy media) separators to clean the coal (Fig. 28). Dense ("heavy") liquid flotation tanks and cyclones are the most common methods of density separation. These devices are designed to separate less dense, "light" coal particles from dense, "heavy" minerals, such as pyrite that contains sulfur, and inorganic rock. However, not all of the inorganic impurities can be removed from coal in this manner. Some mineral matter and minerals filling cell-sized voids in the coal (including tiny sulfur-rich pyrite particles) will remain. In some instances, the fine sizes are further separated into intermediate and still finer size classes and cleaned. It is much more difficult and expensive to clean impurities from the fine size fractions.

Cleaned coal is sized, dewatered, dried, tested to assure that it meets quality parameters, and shipped to the end user. In many areas, different coals may be blended as part of the preparation process. In this way, coals that meet or exceed parameters specified by the end user can be formulated by blending coals of varying qualities.

The coarse-grained waste material produced by processing is called refuse or gob. Gob piles are regulated similar to other non-hazardous solid waste under the Federal Resources Conservation and Recovery Act of 1976, and much of the material is used during reclamation as fill. Fine-grained waste material that remains partly suspended in water is called slurry. On average, 70 to 90 million tons of coal preparation slurry are produced annually in the United States. Shallow ponds called impoundments or slurry ponds hold the slurry (Fig. 29). In the slurry impoundments, the fine waste sediment falls out of suspension and clean water can be recycled through the plant. In 2001, there were 713 fresh-water and coal waste impoundments in the United States.

Fig. 29. Slurry consists of fine particles of sediment (in this case coal and impurities) suspended in water.

Environmental Impacts

The primary environmental concern with transporting coal is

■ Road damage and public safety.

The potential environmental concerns associated with processing coal (and in some cases loading facilities associated with transport or processing coal) include

■ Water quality issues and acidic drainage;
■ Slurry impoundment stability; and
■ Physical disturbances and gob fires.

Fig. 30
KETCHUP LAKE REMEDIATION

Fig. 30. The Pleasant View abandoned (pre-1977) surface mine in western Kentucky was used for disposal of refuse from mines and coal processing. The impoundment was called Ketchup Lake because of discoloration from acidic drainage. Water treatment, dewatering, regrading, liming, and revegetation reclaimed the site.

Road Damage and Public Safety

Road damage can result from the transportation of coal from mines and coal processing facilities. Road damage from coal trucks is a concern in many coal-mining states. Overweight trucks are a contributing factor, especially in the eastern United States where coal trucks sometimes exceed legal limits by more than 50,000 pounds. State enforcement of legal limits ensures public safety and reduces damage and costly repair to busy haul roads. Fatalities have occurred on coal haul roads involving coal trucks, and at railway crossings involving coal trains and thus, public safety is an issue.

Water Quality and Acidic Drainage

Many of the environmental concerns associated with coal waste piles at abandoned (pre-1977) preparation facilities result from pyrite oxidation and production of acidic drainage. Waste piles can have an increased potential for acid formation because pyrite is concentrated in the refuse during coal cleaning. Not all coals and processing wastes have the same potential to produce acidic drainage. In general, acidic drainage from processed wastes is a concern where high-sulfur coals are, or have been, mined and processed. The manner in which acidic drainage forms in refuse piles is similar to that in mines (see Fig. 20, p. 25), and the concerns are similar. Processing facilities use a wide range of practices (active, passive, and engineered structures similar to those discussed in the preceding chapter) to prevent acidic drainage from forming.

To stop acidic drainage in refuse material, water flow through the refuse must be limited. Establishment of vegetative cover can aid in reducing acidic drainage because vegetation stabilizes slopes which limits weathering. Some wetland plants retain metals and stimulate microbial processes that cause metals to precipitate. Properties of the tailings, acidity, and compaction influence the success of revegetation. Methods for limiting acidic drainage include bulk-blending alkaline materials, wetland construction, engineering seals and barriers, or constructing drains and ponds (Fig.30), similar to methods discussed for mines in the previous chapter. The goal of these methods is to prevent water flow through the piles, which limits the formation of leachates and neutralizes any acidic drainage before it can migrate offsite.

Slurry Impoundments

On Feb. 26, 1972, a non-engineered slurry pond impoundment in Buffalo Creek, West Virginia, failed following heavy rains. The slurry rushed downstream, killing 125 people. Following the disaster, the Federal Mine Safety and Health Administration created strict regulations for the permitting and design of new impoundment dams. Impoundments are inspected regularly to look for surface displacement, changes in pore pressure, discharge, and subsurface movement. Groundwater around impoundments is monitored for changes in water level and chemistry. Since these regulations have been in place, no impoundment dams have broken, although in some cases water and slurry bypassed the dam by leaking through underlying abandoned mines (called "breakthroughs") and then discharged into streams.

The most notable recent discharge occurred on Oct. 11, 2000, when an estimated 200 to 250 million gallons of slurry leaked from an impoundment in Martin County, Kentucky. The dam did not fail, but slurry catastrophically leaked into underground mines and flowed into two streams, ultimately affecting 75 miles of waterways (Fig. 31). The National Research Council investigated the spill and proposed new guidelines for all slurry impoundments to avoid future accidental releases. In the wake of this accident, several states developed mine impoundment location and warning systems so that appropriate and

Fig. 31. More than 200 million gallons of slurry discharged from an impoundment in Martin County, Kentucky in October, 2000. The discharge affected 75 miles of waterways downstream.

Fig. 32. The dredge in this slurry pond is removing material from which fine particles of coal will be separated and collected (remined) for use.

timely responses can be made in case of future accidents. In some areas, alternative disposal methods may be possible, such as underground injection and dredging the ponds to recover fine coal particles (Fig. 32). Not all alternatives are practical for all areas.

When an impoundment is closed, it must be reclaimed. The residual ponded slurry water is removed, and the surface is regraded. The regraded surface is covered with topsoil or approved cover materials. Runoff is managed to control erosion and sediment. This surface is then revegetated according to permit requirements (Fig. 33). The reclamation process occurs in phases and regulatory authorities oversee all stages.

Fig. 33. After abandoned slurry ponds and processing refuse are filled and graded, they are covered with agricultural lime or other soil amendments, and then seeded. Limestone channels and other methods for mitigating acidic drainage may be used when acidic leachates are present. The red areas in the top photo are acidic leachate.

Physical Disturbances and Gob Fires

At a preparation facility, the processing plant, piles of unprocessed coal, processed coal, gob, and slurry impoundments are visible disturbances. Likewise, coal piles, and the dust and noise associated with transporting coal are common disturbances at coal-loading facilities. Both active and abandoned (pre-1977), coal waste piles are potential sources of fugitive dust, sediment, and leachates (Fig. 34). Sediment runoff from abandoned, unreclaimed piles has caused clogged streams and acidic drainage.

Some abandoned, unreclaimed gob piles can also be a potential fire hazard. The oxidation of pyrite in a gob pile produces heat, which can lead to spontaneous combustion of coal left in the gob. Some piles burn for decades. The primary physical hazards in such fires are the possibility of the fire spreading, possibly igniting vegetation or structures, as well as noxious smoke and fumes. Fortunately, improved preparation techniques, which leave less coal in the gob to potentially burn, have greatly reduced the occurrence of gob pile fires. Techniques for extinguishing gob fires are similar to those for mine fires discussed in the previous chapter.

Modern preparation and loading facilities are regulated to prevent dust, sediment, and leachate from leaving the site, similar to surface coal mines as previously discussed. Likewise all processing and loading areas must be reclaimed. Much of the gob produced by processing plants is used onsite as fill. In some cases, lime is mixed into gob to decrease surface acidity. Because coal refuse has a low average fertility, fertilizers may be applied to ensure the success of revegetation. Federal surface mining regulations mandate that refuse disposal areas must support self-sustaining vegetation for a minimum of 5 years after closure in the East, and 10 years in western arid climates. During this time, leachate and runoff must meet water quality standards and there must be evidence that water quality will not degrade over time. Aside from standard reclamation practices, some old gob piles can also be remined for the coal they contain, and in some cases, the gob (and fine coal particles it contains) can be used as a feedstock for a new type of power generator, called a fluidized bed combustor (see Chapter 5). Fluidized bed combustion and gasification technologies can both turn gob piles into energy. For example, the Seward Plant, a 521 MW Circulating Fluidized Bed combustion unit in Pennsylvania, started operation in 2004 and consumes 3.5 million tons of gob per year for fuel.

Fig. 34. Gob piles, like this old pile in central Illinois, are composed of waste rock from the processing and mining of coal.

Fig. 35. Today, in the United States, more than 90% is used to produce electric power.

4. Using Coal

Coal combustion accounts for a little more than a third (36%) of global electric power generation, and approximately half of U.S. electric power generation. More than 90% of the coal used in the United States is for electric power generation (Fig. 35). In fact, all but two states, Rhode Island and Vermont, produce some electricity from coal, and seven states produce more than 90% of their electricity from coal. Coal is also used for industrial purposes (6%), to make coke for steel production (2%), and for residential or commercial heating. In addition, coal can be converted to a clean synthesis gas (syngas) through the process of gasification to make chemicals and fuels. The Eastman Chemical Company has gasified coal for more than 20 years to produce carbon monoxide and hydrogen, which are used as chemical building blocks for a wide range of consumer products including Tylenol®, Nutrasweet®, plastics for toothbrush handles, and the celluloid of photographic film.

Power and Heat Generation

In a typical coal-fired plant, the coal is crushed to a fine powder and blown into the furnace (boiler); this method is termed pulverized coal combustion (Fig. 36). Water is pumped through pipes, called a waterwall, surrounding

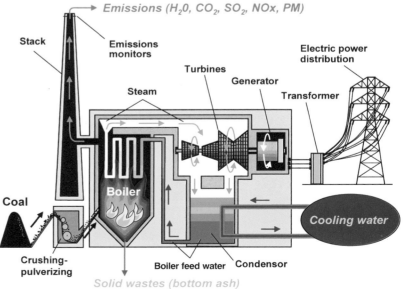

Fig. 36. Diagram of electricity production in a typical pulverized coal-fired steam plant. Various technologies discussed in this chapter may be added to the boiler or between the boiler and stack to limit emissions.

the inside walls of the furnace; the heat generated from the combustion of the coal converts the water to steam. The high-pressure steam turns giant turbines that drive a generator (a magnet that rotates inside a jacket of copper wires) to produce electricity. Other methods for producing electricity from coal, such as gasification, are discussed in the next chapter.

In traditional coal-fired power plants it takes on average one ton (2,000 pounds) of coal to produce 2,500 kilowatt hours (kWh) of electricity. The actual amount of coal needed varies with coal heating value and power plant design. An average U.S. home uses between 900 and 1000 kWh/month, equating to 720 to 1000 pounds of coal a month or four to six tons of coal per year. Another way to look at our coal use is on a per capita basis. In 2004, we produced 1.1 billion short tons of coal, or 3.7 short tons of coal per person. Ninety percent of that, or 3.3 short tons per person was used to generate electricity!

Coal is also used to heat and power foundries, cement plants, and other industrial and manufacturing facilities. Steel mills use coal to make a carbon-rich material called "coke", which serves as a heat source and oxygen-reducing agent for smelting iron to make steel. In addition, the coking process yields numerous useful byproducts. Coal tar, light oils, and ammonia are among the many coal-derived materials, which are used as chemical feed stocks to make a variety of chemicals. In the past, coal was widely used for home heating, but now this use accounts for less then 1% of annual U.S. coal consumption.

Impacts of Coal Use

The combustion of coal in pulverized coal power plants and in other industries, occurs at very high temperatures (>2000° F). When burned, many of the elements in coal are converted to gaseous or solid "oxides." Airborne emissions of these oxides, along with solid byproducts are the principle environmental concerns with coal combustion. The amount and type of coal used, the size and type of electric-generating technology used at the plant, and the area in which the power plant is located determine the types of environmental concerns that may be associated with airborne emissions and solid byproducts. The impacts of the various emissions from coal-fired power plants include

- Sulfur oxides — acid rain;
- Nitrogen oxides — acid rain, ozone, and smog;
- Particulates — haze;
- Mercury — health effects;
- Carbon dioxide — climate change; and
- Solid byproducts — waste disposal issues.

Sulfur Oxides and Acid Rain

Electric power generation currently accounts for two thirds (67%) of U.S. sulfur dioxide (SO_2) emissions. Sulfur dioxide forms when sulfur in the coal combines with oxygen in the furnace. In the atmosphere, SO_2 can react with water vapor to form sulfurous acid (H_2SO_3) which oxidizes to sulfuric acid (H_2SO_4), components of "acid rain" (Fig. 37). Sulfates in the atmosphere, both wet and dry, contribute to sulfur deposition. Crop damage, forest degradation, impaired visibility, chemical weathering of building stones and monuments, increased acidity of lakes and streams, and increased human health risks from asthma and bronchitis have been attributed to acid rain. In the United States, sulfate precipitation is greatest in the Northeast, which has been attributed to coal-fired power plants in the Midwest and Northeast (Fig. 38).

Because of these environmental issues, the federal government passed the Clean Air Act Amendments, which mandated the lowering of SO_2 emissions from power plants. The Acid Rain Program (Title IV of the 1990 Clean Air Act) called for SO_2 and NO_x reductions in two phases, primarily through a cap and trade program that allowed utilities marketable allowances and choice of compliance methods. Marketable allowances mean that utilities can buy and trade emissions credits to meet regional emissions goals rather than enforcing one limit on all plants at the same time. Thus, utilities may meet new standards through a variety of mechanisms so that total regional emissions are lowered without significant economic impact to the consumer. Older, coal-burning power plants were exempted from the ruling (the "grandfather" clause) with the idea that they would ultimately be replaced by newer plants with advanced emissions control technologies.

The result of these regulations has been a nationwide decrease in SO_2 emissions of nearly 40% from 1980 levels. It is important to note that this decrease was achieved while coal production and use

ACID RAIN

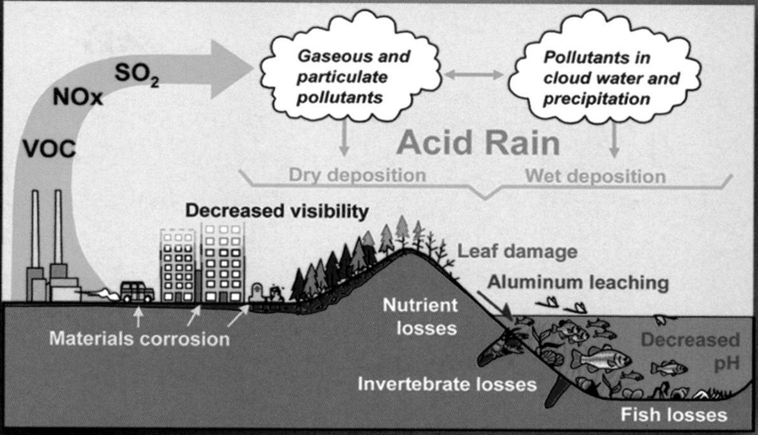

Fig. 37. To reduce acid rain, restrictions were placed on sulfate emissions at power plants.

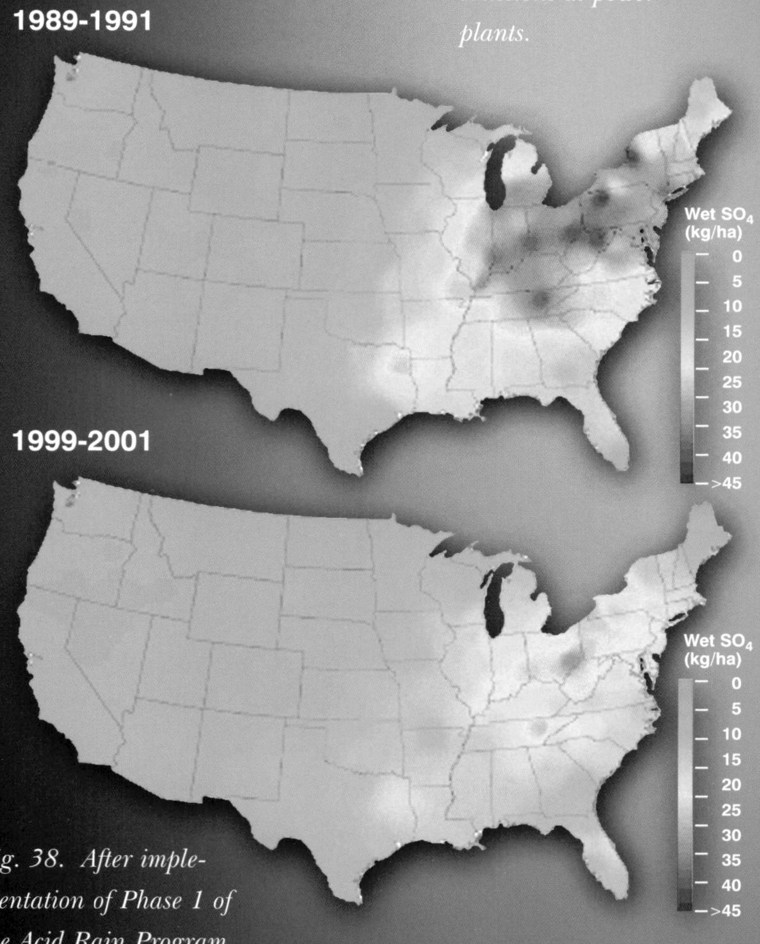

Fig. 38. After implementation of Phase 1 of the Acid Rain Program, deposition of wet sulfate decreased, especially in the northeastern states.

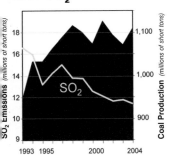

Fig. 39. SO_2 emissions decreased following enactment of the Clean Air Act Amendments while coal production and use increased.

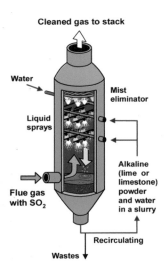

Fig. 40. Diagram of a "wet" spray tower scrubber showing how SO_2 is removed from the flue gas.

increased (Fig. 39). Further reductions will be required by the Clean Air Interstate Rule, which was issued in 2005. The rule permanently caps emissions of sulfur dioxide and nitrogen oxides in 28 eastern states (where concentrations are the greatest). The regulation is expected to further reduce sulfur dioxide emissions more than 70% from 2003 levels by 2015.

Several methods are used to decrease sulfur emissions:

- Switching to low-sulfur coals;
- Increased processing of higher sulfur coals;
- Retiring old (less efficient) power plants; and
- Using clean-coal technologies.

The most common clean-coal technology used to capture SO_2 is flue-gas desulfurization, commonly called "scrubbing." Scrubbers have been required on conventional coal power plants built since 1978, and many have been added to grandfathered plants built prior to 1978. Scrubbers work by injecting either dry or slurried (wet) alkaline material into the path of the flue gas leaving the furnace (Fig. 40). A reaction occurs that converts gaseous SO_2 in the flue gas to wet byproducts that oxidize into solid calcium sulfate (gypsum), which falls to the bottom of the scrubber where it is removed as a solid or slurry or is collected as dust in a baghouse or electrostatic precipitator. Both wet and dry scrubbers can remove more than 90% of the SO_2 in flue gas. The gypsum byproduct can be used to make wallboard for the construction industry. In fact, some of the largest wallboard plants in the United States use synthetic gypsum from scrubbers.

The U.S. Department of Energy is demonstrating Fluidized Bed Combustion and Integrated Gasification Combined Cycle technology in partnership with several power plants in the United States. These technologies, which will result in substantial decreases in sulfur dioxide emissions (in combination with other emissions), are discussed on pp 55-56.

NO_x, Acid Rain, Smog, and Ozone

Nitrogen oxides (NO_x) are a group of reactive gases that contain nitrogen and oxygen. NO_x emissions from coal-fired power plants are an environmental concern because they contribute to the formation of acid rain by combining with atmospheric water to form nitric acid. NO_x emissions also contribute to the formation of ground-level ozone (O_3). Ground-level ozone is an important ingredient of urban smog, which is a respiratory irritant. Increased ground-level ozone can also lead to reduction in agricultural crop and commercial forest yields.

Electric utilities were responsible for 22% of NO_x emissions, second behind the transportation sector (cars, trucks, etc.), which produced 55% of NO_x emissions in 2003. In order to decrease NO_x emissions, Phase 1 of the 1990 Clean Air Act Amendments mandated decreases in NO_x emissions for 239 coal-fired power plants built since 1978. Older grandfathered plants were exempted from the ruling.

In 2003, further reductions in NO_x emissions were required in the eastern United States (where NO_x concentrations are the greatest) as part of the Ozone Transport Rule. In 2005, the U.S. Environmental Protection Agency issued the Clean Air Interstate Rule, which set caps for nitrogen oxides (and other emissions) in the eastern United States. The regulation is expected to further reduce NO_x emissions 60% from 2003 levels by 2015.

To meet NO_x levels for Phase 1 of the Clean Air Act, more than half of the affected utilities chose to reduce NO_x emissions with advanced burner technology. Rather than capturing emissions after combustion, as with SO_2 scrubbers, advanced burner technology reduces NO_x by using staged combustion; the air is mixed with the fuel in stages, which lowers the combustion temperature and reduces the concentration of oxygen, both of which reduce NO_x formation. Advanced burners, such as Low-NO_x Burners, regulate the fuel/air mixture entering the furnace (Fig. 41), and can achieve 35 to 55% decrease in NO_x emissions. Other technologies that are being used to reduce NO_x emissions include

- Selective Catalytic Reduction (SCR);
- Selective non-catalytic reduction;
- Fluidized bed combustors (see p. 55);
- Overfire air combustion;
- Fuel-reburning; and
- Integrated Gasification Combined Cycle (IGCC) (see p. 66).

To meet more stringent NO_x emission limits, SCR has become the technology of choice for many conventional pulverized coal plants. IGGCs also have demonstrated very low NO_x emissions.

Particulate Emissions and Haze

Particulate matter is solid particles or liquid particles in the air. Such material is an environmental concern because it contributes to smog. Although particulate matter is formed from a wide variety of natural and man-made sources, emissions from fossil-fuel power plants are a particular concern. The U.S. National Park Service has attributed decreased visibility at Great Smokey Mountains, Shenandoah, Mammoth Cave, and Grand Canyon National parks to smog produced upwind by fossil-fuel power plants, exhaust from automobiles, and other sources. Increased particulate matter is also a health concern because it may contribute to respiratory illnesses.

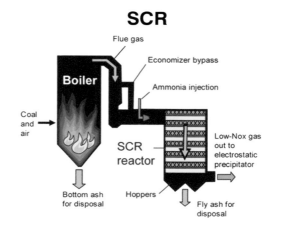

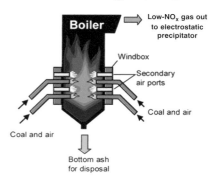

Fig. 41. NO_x emissions can be reduced in the boiler with technology like a Low-NO_x Burner, or between the boiler and the stacks with technology like SCR.

Filter baghouse

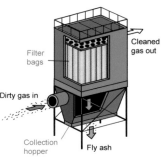

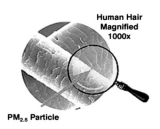

Fig. 43. $PM_{2.5}$ size compared to the width of a human hair.

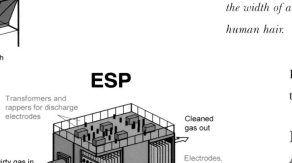

Fig. 42. Diagrams of fabric filters and electrostatic precipitators (ESPs), which are used to capture particulate matter.

Fig. 44. Mercury increases in concentration up the food chain through bioaccumulation.

The two principal technologies for limiting particulate matter in coal-fired power plants are electrostatic precipitators and fabric filters, sometimes called "baghouses" (Fig. 42). Electrostatic precipitators work by charging the particles in the flue gas stream across a series of wires or plates. The particles are attracted to plates that carry the opposite charge. The particles (ash) are then removed from the plates and collected in hoppers for disposal. Fabric filters work like an air filter in your home furnace. Both technologies are very efficient, and typically capture more than 99% of the particulate matter in the flue gas stream.

Electrostatic precipitators and fabric filters have generally been installed to prevent particulates from escaping into the atmosphere. Particulates, with diameters of 10 microns or less (a micron, or micrometer, is 0.00004 inches and 2,500 10-micron particles would fill one inch), are referred to as PM_{10}. Smaller particles — those with diameters of 2.5 microns or less are called $PM_{2.5}$ (Fig. 43). The smaller a particle is, the more difficult it is to remove.

Mercury and Hazardous Air Pollutants

The U.S. Environmental Protection Agency listed 189 substances as hazardous air pollutants (HAPs) in Title III of the Clean Air Act Amendments of 1990. Of these, 15 occur in coals: antimony, arsenic, beryllium, cadmium, chlorine, chromium, cobalt, fluorine, lead, manganese, mercury, nickel, selenium, thorium, and uranium. Not all coals contain all 15 of these HAPs and concentrations of the HAPs vary significantly among different coal beds. Most HAPs occur as "trace" elements in very tiny amounts, typically measured in parts per million. The U.S. Environmental Protection Agency studied the amounts of hazardous air pollutants emitted from coal-fired power plants and concluded that emissions of HAPs in coal, with the exception of mercury, do not pose a threat to human health.

The potential health hazard from mercury is that one form, called methylmercury, can accumulate in invertebrates and fish and is a potent neurotoxin. Through the process of bioaccumulation, the concentration of mercury increases up the food chain—when one fish eats another, any mercury in the fish that was eaten becomes part of the survivor (Fig. 44). For certain oceanic fish, and in rivers, ponds, and lakes where high levels of mercury are a concern, advisories are issued by regulatory agencies.

Power plants are currently the largest emitters of mercury into the atmosphere in the United States. The emitted mercury is non-hazardous elemental mercury, but it can be transformed into methylmercury if deposited in an aquatic environment. Total annual emissions of mercury from power plants in the United States are estimated to be 48 tons. Although that amount is less than 1 percent of global mercury emissions, it accounts for a third of our country's total anthropogenic (man-made) mercury emissions (Fig. 45).

In order to decrease the amount of mercury being deposited in our nation's lakes and streams, the U.S. Environmental Protection Agency issued the Clean Air Mercury Rule in March, 2005. This regulation makes the United States the first country in the world to regulate mercury emissions from power plants. The rule targets mercury emissions from coal-fired power plants through a market-based cap-and-trade program (similar to that used successfully for reductions of sulfur emissions). Phase 1 sets a cap of 38 tons of mercury in 2010, which can be met with projected installation of technologies designed for reduction of sulfur dioxide, nitrogen oxides, and particulate matter.

Phase 2 of the Clean Air Mercury Rule caps total national mercury emissions at 15 tons in 2018, which will require the installation of mercury-specific control technologies. One of the most promising technologies is called activated carbon injection. This process works by injecting powdered "activated carbon" into the flue gas. Activated carbon is carbon that has been treated to alter its surface properties so that it will act as a sorbent and bind to mercury. The mercury-containing particles can then be captured by a particulate control device. Mercury removal rates of more than 90% have been achieved using this technology in large-scale field demonstrations on coal-fired power plants.

Carbon Dioxide

When coal burns, carbon in the coal oxidizes forming carbon dioxide (CO_2), a greenhouse gas. Greenhouse gases are gases capable of absorbing infrared radiation as it is reflected from the Earth back towards space, trapping heat in the atmosphere (Fig. 46). Many naturally occurring gases exhibit "greenhouse" properties. Water is the most abundant greenhouse gas. The concern in recent years has been that atmospheric concentrations of several important greenhouse gases, such as carbon dioxide, methane, and nitrous oxide, have increased since large-scale industrialization began 150 years ago. These increases are thought to be contributing to global climate change.

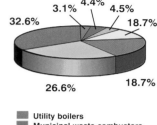

Fig. 45. Power plants account for approximately one third of U.S. mercury emissions.

Fig. 46. Carbon dioxide is a greenhouse gas. Greenhouse gases are produced naturally and by man, but increased man-made emissions may influence climate.

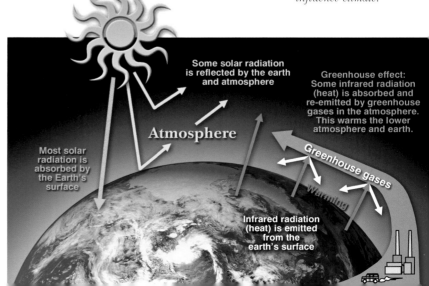

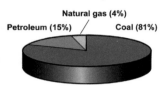

CO₂ by Fuel Type in the Electricity Sector

Natural gas (4%)
Petroleum (15%)
Coal (81%)

Fig. 47. Coal's combustion is responsible for most of the CO_2 produced by electric power plants.

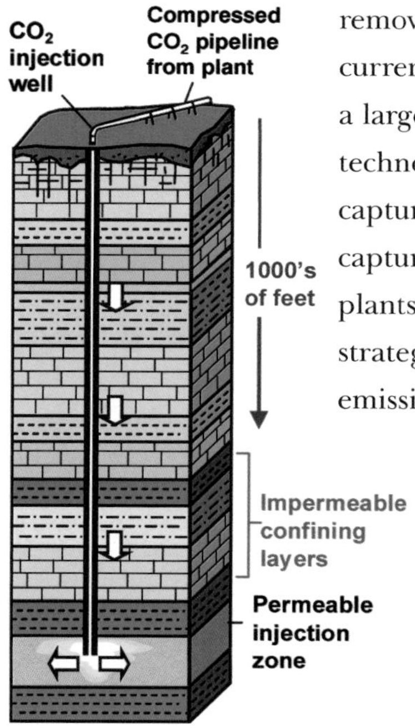

Fig. 48. Carbon dioxide can be injected into rock units deep underground.

Because coal contains more carbon per unit of energy than other fossil fuels, it will produce more carbon dioxide per unit of electric power production than petroleum or natural gas. Electrical utilities, being the largest consumers of coal, are the largest sources of CO_2 emissions from coal (Fig. 47).

In December 1997 at a meeting in Kyoto, Japan, many of the developed nations agreed to limit their greenhouse gas emissions, relative to the levels emitted in 1990. The United States, the largest producer of CO_2, did not ratify the treaty, citing potential harm to our economy. China and India, the second and third largest producers of CO_2, and many developing countries were excused from adherence to the CO_2 emission limits. Both China and India have large coal resources and use coal to generate electrical power, as well as for industrial and residential uses.

Although scrubbers exist that can remove CO_2 from flue gas, they are not currently economically feasible. There is a large global effort to improve existing technology and develop new processes for capturing CO_2 from flue gas, so that CO_2 capture from pulverized coal-fired power plants may be economic in the future. Some strategies that might be used to reduce CO_2 emissions include (1) changing fuel sources to less carbon-rich fuels, such as natural gas, nuclear energy, solar, wind, and bioenergy, (2) increasing the efficiency of electrical power production so less CO_2 is emitted per unit energy produced, (3) retrofitting old plants with new, more efficient technology, and (4) developing ways to capture and sequester (permanently store) CO_2 to prevent its emissions to the atmosphere. Three general types of storage options are being investigated.

Terrestrial sequestration involves optimizing agricultural processes, soil reclamation with coal combustion byproducts, and increased forestry (tree planting) to offset greenhouse emissions. Oceanic sequestration would involve the injection of CO_2 into ocean-bottom sediments or ice-like gas hydrates, but there are concerns about impacts to the ocean ecosystem. Geologic sequestration involves pumping captured CO_2 gas under pressure into a suitable rock layer deep under the ground (Fig. 48).

For geologic sequestration, depleted oil and gas fields, unmineable coal beds, organic-rich shales, and saline water-bearing formations have all been identified as potential repositories for CO_2. In depleted oil and gas fields, the CO_2 could be used to enhance oil recovery, which would provide an economic incentive for sequestration.

In enhanced (or secondary) recovery CO_2 is injected into an oil-bearing reservoir and displaces or mixes with the oil it contacts in the reservoir, reducing its viscosity so that it can be more readily recovered. Carbon dioxide is already used for secondary recovery in Texas and other parts of the United States so that the technology is available and tested. Approximately 5,000 tons/day of nearly pure CO_2 produced at the Great Plains Coal Gasification Plant in North Dakota are shipped through a 204-mile pipeline to the Weyburn oil field in Canada. This project hopes to add another 20 years and recover as much as 130 million barrels of oil from a field that might otherwise have been abandoned.

In unmineable coal beds, injected CO_2 can displace methane adsorbed (bound) to the coal surface. The methane could be produced as a secondary energy resource. In this way coal beds could be used to produce a useful gas, while sequestering a waste gas.

The U.S. Department of Energy has established a Carbon Sequestration Program to develop advanced technologies to reduce greenhouse gas emissions, including carbon dioxide. To help reach this goal, the Department of Energy has established seven Regional Carbon Sequestration Partnerships, consisting of governmental, industrial, educational, and other entities, to determine the most suitable technologies, regulations, and infrastructure needs for carbon capture, storage, and sequestration in different parts of the United States. A suite of commercially ready sequestration technologies and options are being investigated because no one method or option will suit all needs.

Solid Waste Byproducts

The combustion of coal by electrical utilities produces several solid waste byproducts, referred to as coal combustion byproducts or coal combustion wastes. These materials include fly ash, bottom ash, and flue-gas desulfurization byproducts from conventional coal-fired combustion, as well as slag and ash from gasification processes.

Increased sulfate and trace element concentrations from leachate are a potential concern because elements such as barium, boron, iron, manganese, and selenium can be concentrated in fly ash and might be mobilized under certain conditions in leaching waters. Electric utilities monitor the pH of ash disposal areas specifically for this reason. Although a potential concern, research suggests that less than 1% of coal combustion byproduct wastes have potential to leach hazardous elements. Careful design of impoundments and landfills, with placement in areas where the geology and hydrology are favorable for containment of any potential leachates, are key elements to preventing or limiting future environmental impacts.

EPA has encouraged the use of coal combustion byproducts to reduce solid wastes. Byproducts, such as fly ash or scrubber waste have been used in the making of construction materials such as wall board, concrete block, and bricks, where this can be done in an environmentally safe manner. Currently, about one third of the coal ash and just over one fourth of the scrubber waste produced in coal-fired power plants are recycled for commercially beneficial uses (Fig. 49).

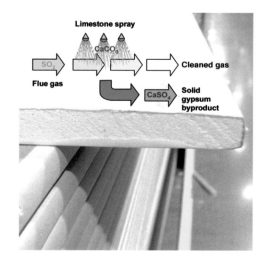

Fig. 49. Wallboard is made from gypsum, which is a byproduct of scrubber waste.

Fig. 50. Coal is our most plentiful energy resource. It is imperative that society develop the appropriate balance of policies for maximizing the use of our country's resources, meeting energy needs, and providing a healthy environment.

5 Providing for the Future

Our country requires a lot of energy, and coal is our most plentiful energy resource (Fig. 50). As such, coal will remain the backbone of the nation's electrical power generation for the foreseeable future. The challenge is to mine, transport, and use coal in an environmentally acceptable manner. Meeting this challenge involves

- Comprehensive understanding of the composition (quality), quantity, and distribution of our country's coal resources and reserves;
- Mining, transportation, and processing practices that minimize disturbances to the land and pollution of surface and groundwater;
- Sound reclamation methods that restore mined lands and allow beneficial post-mine land uses (Fig. 51);
- Effective regulations based on scientific data (Fig. 52);
- Increased use of clean-coal technologies to decrease harmful emissions; and
- Continued research and development of new technologies that allow coal to be used in an enviromentally acceptable way.

Fig. 51. This wildlife area in central Illinois was a large abandoned surface mine. The property, including more than 200 lakes, has revegetated naturally and is used for hunting, fishing, and recreation.

Support for Technology Development

As the environmental impacts of past coal mining, processing, and utilization have been recognized, scientists and engineers have worked

Fig. 52

RULES & REGULATIONS

The National Environmental Policy Act (NEPA)
The National Environmental Policy Act, passed in 1969, established the basic environmental policies for the United States and provided a process to handle decisions regarding mine development on federal lands. The Council of Environmental Quality establishes NEPA guidelines. The Department of Interior administers NEPA policies related to mining and construction on public domain lands. The Department of Agriculture administers NEPA on National Forest System lands and Grasslands.

Mine Safety and Health Act
In 1969, the Federal Coal Mine Health and Safety Act (amended in 1977 and renamed the Mine Safety and Health Act) was passed to improve working conditions at mines and address the health issues of coal, stone aggregate, and metal miners. The Act establishes mandatory health and safety standards and empowers inspectors to enforce compliance by the operators of the mine. The Mine Safety and Health Administration (MSHA) administers the programs.

Clean Air Act (CAA)
The Clean Air Act, passed in 1970, provides strict requirements for preventing and controlling major air pollutants that may be hazardous to human health or natural resources. The CAA authorizes the U.S. Environmental Protection Agency to establish National Ambient Air Quality Standards to protect public health and the environment. The CAA was amended in 1977 primarily to set new compliance deadlines. The 1990 amendments to the CAA addressed several additional issues, including those posed by the impact of acid rain.

Clean Water Act (CWA)
Initially referred to as the Federal Water Pollution Control Act of 1972 and amended in 1977, the Clean Water Act authorizes regulations that cover discharges of pollutants into the waters of the United States. It specifically sets guidelines for coal mine-water discharges. This act is also the focus of attention in the mountaintop mining controversy in the eastern United States; specifically, the regulation against dumping within 100-feet of a stream.

Surface Mine Control and Reclamation Act (SMCRA)
In 1977, Congress enacted the Surface Mine Control and Reclamation Act, which gave individual states with established federal-approved enforcement programs the primary responsibility for enforcing surface coal mining regulations in their jurisdictions. In areas where the program does not exist, SMCRA is implemented by the Federal Office of Surface Mining in the Department of Interior. The Act sets performance standards for mining operations to protect the environment and guarantee reclamation.

SMCRA requires that coal companies obtain mining permits from pertinent local, state and federal regulatory agencies prior to mining. To assure that lands being mined will be restored to approximately the same physical contour and state of productivity equal to or better than pre-mining conditions, regulations require companies to post reclamation bonds, which are not returned to the company until all aspects of the permit have been met.

to understand the underlying mechanisms of the impacts and develop mitigation strategies. Many innovations have been developed from cooperative research between federal agencies such as the U.S. Department of Energy, U.S. Environmental Protection Agency, and the Office of Surface Mining, and private industry (mining companies, power plants, steel mills, cement plants, etc.), state agencies, and universities. This research is ongoing.

The U.S. Department of Energy's Clean Coal Technology Program is a partnership between the federal government, industry, and universities; the objective is to develop, test, and demonstrate technologies at commercial scale that utilize coal for energy production. Among other achievements, the program helped demonstrate and lower the cost of effective scrubbers for sulfur dioxide emissions. Numerous promising technologies have been developed and tested including Fluidized Bed Combustion and Integrated Gasification Combined Cycle plants. Another important cooperative program is the FutureGen initiative. FutureGen will not only limit emissions, but will also produce a wide range of products, including liquid fuels, chemical feedstocks, and hydrogen.

Future Electricity from Clean Coal Technologies

Fluidized Bed Combustion

Fluidized bed combustion is a type of clean coal technology that is being used in electrical power generation because of its increased efficiency and decreased emissions (Fig. 53). This combustor was developed largely through the technology program of the U.S. Department of Energy's Office of Fossil Energy (and its predecessors). 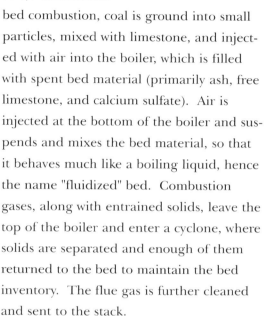 In fluidized bed combustion, coal is ground into small particles, mixed with limestone, and injected with air into the boiler, which is filled with spent bed material (primarily ash, free limestone, and calcium sulfate). Air is injected at the bottom of the boiler and suspends and mixes the bed material, so that it behaves much like a boiling liquid, hence the name "fluidized" bed. Combustion gases, along with entrained solids, leave the top of the boiler and enter a cyclone, where solids are separated and enough of them returned to the bed to maintain the bed inventory. The flue gas is further cleaned and sent to the stack.

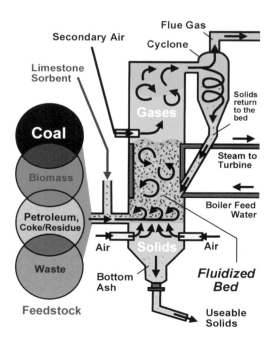

Fig. 53. Diagram of a Fluidized Bed Combustor unit.

In a fluidized bed combustor, sulfur in the coal reacts with lime from the limestone to form calcium sulfate, resulting in more than 90% sulfur capture. Because of the complete and rapid mixing in the bed, boiler temperature is lower (about 1,500°F vs. 3,000°F for a pulverized coal boiler) which decreases NO_X production. Steam, generated in tubes embedded in the fluidized bed, is sent to a steam turbine to generate electricity.

Aside from the benefits of reduced emissions, another advantage of fluidized bed combustors is that they can handle a wide range of carbon-based feedstocks, from coal to municipal waste. Currently, 27 fluidized bed combustors beneficially use 12.8 million tons of gob in the United States (mostly in Pennsylvania and West Virginia) annually. More information about this technology can be found at the U.S. Department of Energy website, www.energy.gov/.

Gasification Technology

The next generation of electric power generation using coal will probably use gasification technology. Gasification technology allows for the possibility of combining electric power generation with the production of synfuels and chemical products (Fig. 54). In an Integrated Gasification Combined Cycle (IGCC), a gasifier uses intense heat and pressure to convert coal and other solid carbon-based feedstocks (petroleum coke, refining liquids, biomass, solid waste, tires, etc.) into a synthetic gas, also called syngas.

In a gasifier, coal is fed into a chamber together with an amount of oxygen (or air) that is insufficient to achieve complete combustion and steam at high temperature and moderate pressure. Under these conditions, the coal is gasified (rather than combusted) to produce a mixture of gases, including carbon monoxide, hydrogen, methane, carbon dioxide, hydrogen sulfide,

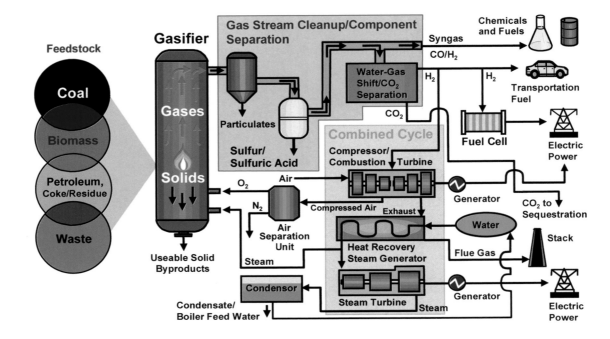

Fig. 54. Diagram of a Integrated Gasification Combined Cycle.

and ammonia. Because of the reducing conditions, that is, the presence of hydrogen in the gasifier, sulfur and nitrogen impurities are bound to hydrogen to form hydrides, rather than to oxygen to form oxides, as occurs in a pulverized coal-fired boiler. The fuel gas leaving the gasifier is cooled, cleaned of particulates and ammonia and hydrogen sulfide (hydrides), and sent to a gas turbine. The ammonia can be recovered and sold as a fertilizer, and the hydrogen sulfide can be converted to sulfur or sulfuric acid for chemicals and other uses.

More than 99% sulfur removal is possible with gasification technology. Because gasifiers breakdown carbon-rich fuels in a reducing (low oxygen) environment, NO_x is significantly reduced. IGCCs use a combination of cyclones and ceramic or metal-filter elements to remove particulate emissions, achieving removal efficiencies of 99.9% or greater. Likewise, in excess of 95% of mercury can be removed from syngas in IGCCs using activated carbon beds (see p. 48 in Mercury section). Mercury removal from syngas generated from coal gasification has been demonstrated for more than 20 years at the Eastman Chemical Company.

In an integrated combined cycle gasification (IGCC) plant, the syngas produced by the gasifier is combusted in a combustion turbine. The turbine drives an electric generator to produce electrical energy. Heat from the turbine exhaust is used to generate steam, which is then used to drive a steam-turbine; hence two turbines for a combined cycle. Some of the steam is also used for the gasifier. Because both a steam and combustion turbine are used to generate electric power, the efficiency of the plant is increased. Because efficiency is increased, less CO_2 is produced per kilowatt hour than a standard pulverized coal combustion plant.

In addition to reducing the relative amount of CO_2 produced, the carbon monoxide (CO) in an IGCC's fuel gas can be converted to carbon dioxide (CO_2) and hydrogen. The hydrogen can be used for refinery applications. As fuel cell technology and efficiency increases, it is also possible that the hydrogen could be used to run hydrogen fuel cells for electric power generation, which would further increase the efficiency of these plants. The CO_2 can be recovered and injected into geologic formations deep underground for permanent storage or use for enhanced oil and gas recovery, if desired, permitting the production of power from coal with very low greenhouse gas emissions. Also, recovering CO_2 in an IGCC system should be much cheaper than in a pulverized coal plant, which is why gasification technology is planned for the power plants of the future, including the FutureGen initiative.

FutureGen

FutureGen is a 10-year, $1 billion, DOE-industry partnership-initiative to build the world's first zero-emissions power plant. When operational, the prototype will be the cleanest fossil fuel-powered plant in the world. FutureGen will be able to burn coal or any carbon-based feed stock with near-zero emissions. Gasification technology will be integrated with combined cycle electricity generation and the sequestration of carbon dioxide emissions. The plant will establish the technical and economic basis for producing electricity and hydrogen from our nation's vast coal resources, while limiting the emissions of environmental concern discussed in the preceding chapter. Because hydrogen is produced as a product of gasification, the plant will provide a cost-effective way to produce hydrogen for use in transportation, fuel cells, and other applications. More information on FutureGen, gasification, and sequestration can be found at the U.S. Department of Energy's website, www.energy.gov/.

Liquid Fuels from Coal

In the near-future, coal may also be used to generate liquid fuel that is sulfur, nitrogen, and mercury free. Coal-derived liquid fuel is an excellent diesel fuel and can be used directly or blended with refinery streams to produce diesel fuel with reduced emissions. Coal is already used to make liquid fuels in South Africa. Several countries, including China and Australia, are investing in coal-to-liquids technology to meet their rising energy needs. This technology is not new. The process for converting coal to a liquid fuel, uses a gasifier to convert the coal to a syngas, and then the syngas is converted to a liquid through the Fischer-Tropsch (FT) process. FT synfuels were developed in the 1920's in Germany, and helped power the German war machine during World War II. Research continued in many countries following the war, but low oil prices kept the process from being economical in most countries. Higher petroleum prices and increased energy needs have renewed interest in FT fuels, including the possibility that U.S. coal-sourced FT fuels would be used by the U.S. military.

The Future of Coal

Because coal is our country's most abundant energy resource, it will remain important to our energy mix. We need to understand the environmental impacts of coal use, and prevent or mitigate those impacts while still providing secure and inexpensive energy. To some it may seem that our use of coal is at odds with environmental protection. The science and economics behind various sides of the arguments that arise between the energy sector and environmentalists are complicated, and, unfortunately, sometimes emotional. In some cases, incomplete understanding of the science and economics further complicates these issues. Yet reasoned approaches by all concerned can provide balanced solutions to our country's energy needs, while still protecting our environment.

Prevention and mitigation of environmental problems regarding coal use can be achieved through responsible mining, preparation, and utilization, in concert with oversight by industry, citizens groups, and state and federal regulatory agencies. These issues cannot be solved by any one sector of our society. Government should provide incentives and initiatives to help energy and utilization industries implement new environmental technologies, while keeping the costs of energy low for consumers. Individual citizens can help by reasonably limiting their energy use to what is necessary and by recycling materials. Even if paper, plastic, and other recyclables do not directly use coal in their manufacture, the power to manufacture many everyday items comes from electricity generated in coal-fired power plants. In the end, the amount of coal mined is proportional to our energy demands, which everyone helps determine.

New technologies and mitigation strategies continue to be developed to meet higher environmental standards. For example, research continues into finding economic and environmentally safe techniques for gasifying coal in the ground, called in situ gasification. In some areas this would reduce the need to mine coal, and therefore many of the environmental impacts associated with mining and processing coal. Research also continues into finding alternative uses for coal combustion byproducts, which would decrease the amount of solid wastes. Research into technologies and strategies that will decrease anthropogenic carbon emissions are being investigated and tested. As old coal-fired plants are retired, plants with new gasification technologies will be built that gasify, rather than burn coal, producing fewer emissions and operating more efficiently. Continued research into carbon sequestration and FutureGen will result in power plants that can use coal with near-zero emissions. The U.S. Government dedicated more than $9 billion for near-term coal-related energy projects and research in the 2005 Energy Bill. This funding shows the strong commitment the government has to our nation's use of domestic coal for energy purposes. Because coal's use as a fuel will likely continue and even grow, it is imperative that society develop the appropriate balance of policies for maximizing the use of our country's resources, meeting energy needs, and providing a healthy environment both here and abroad.

References and Web Resources

*N*umerous references were used in the compilation of this manuscript. A complete listing by subject can be found at the Kentucky Geological Survey website, www.uky.edu/KGS under coal. Many of the references are available from the Internet, including those from federal agencies such as the U.S. Department of Energy, Environmental Protection Agency, and Office of Surface Mining. State geological surveys, www.stategeologists.org/, environmental agencies, and organizations such as the American Geological Institute, www.agiweb.org, commonly provide educational resources. Organizations with educational resources on the Internet about coal and the environment are highlighted here.

Alabama Department of Industrial Relations-Mining and Reclamation
dir.alabama.gov/mr/

Alaska Department of Natural Resources /Division of Mining Land and Water
www.dnr.state.ak.us/mlw/mining/coal/index.htm

American Coal Foundation
www.teachcoal.org/

Colorado Division of Minerals and Geology
mining.state.co.us/

Colorado Geological Survey
geosurvey.state.co.us/

Energy Information Administration, U.S. Dept. of Energy
www.eia.doe.gov/

Fossil Fuels, U.S. Dept. of Energy
www.energy.gov/energysources/fossilfuels.htm

Illinois State Geological Survey
www.isgs.uiuc.edu/isgshome.html

Illinois Department of Natural Resources
www.dnr.state.il.us/mines/lrd/welcome.htm

Indiana Department of Natural Resources
www.in.gov/dnr/

Indiana Geological Survey
http://igs.indiana.edu/

Kentucky Coal Education
www.coaleducation.org/

Kentucky Division of Abandoned Mine Lands
www.aml.ky.gov/

Kentucky Geological Survey
www.uky.edu/KGS/home.htm

Mineral Information Institute
www.mii.org/

Mine Safety and Health Administration, U.S. Dept. of Labor
www.msha.gov/

Montana Bureau of Mines and Geology
www.mbmg.mtech.edu/

National Association of Abandoned Mine Land Programs
www. onenet/~naamlp/

National Energy Technology Laboratory, U.S. Dept. of Energy
www.netl.doe.gov/

North Dakota Geological Survey
www.ndsu.nodak.edu/instruct/schwert/ndgs/nd_coal.htm

Office of Surface Mining, U.S. Dept. of Interior
www.osmre.gov/osm.htm

Ohio Department of Natural Resources /Minerals Resources Management
www.dnr.state.oh.us/mineral/

Pennsylvania Department of Environmental Protection, Bureau of Abandoned Mine Reclamation
www.dep.state.pa.us/dep/deputate/minres/bamr/bamr.htm

Pennsylvania Geological Survey
www.dcnr.state.pa.us/topogeo/indexbig.htm

Texas Bureau of Economic Geology
www.beg.utexas.edu/

Texas Railroad Commission
www.rrc.state.tx.us/

U.S. Environmental Protection Agency
www.epa.gov/

U.S. Department of Energy
www.energy.gov/

U.S. Geological Survey
www.usgs.gov

Utah Geological Survey
www.geology.utah.gov

Virginia Cooperative Extension, Natural Resources and Environmental Management
www.ext.vt.edu/cgi-bin/WebObjects/Docs.woa/wa/getcat?cat=ir-nrem-mr

Virginia Department of Mines, Minerals, and Energy
www.mme.state.va.us/Dmlr/default.htm

Virginia Department of Environmental Quality
www.deq.state.va.us/

West Virginia Department of Environmental Protection
www.dep.state.wv.us/

West Virginia Geological and Economic Survey
www.wvgs.wvnet.edu/

West Virginia Water Research Institute, National Mine Land Reclamation Center
wvwri.nrcce.wvu.edu/programs/nmlrc/index.cfm

Wyoming Department of Environmental Quality
http://deq.state.wy.us/

Wyoming State Geological Survey
www.wsgs.uwyo.edu/

Wyoming Mining Association
www.wma-minelife.com/coal/coalhome.html

Online Glossaries of Coal and Environmental Terms

California Air Resources Board—Glossary of air pollution terms
http://www.arb.ca.gov/html/gloss.htm

Kentucky Coal Education—Coal mining terms
www.coaleducation.org/glossary.htm

National Academies Press—Glossary for coal waste impoundments
http://www.nap.edu/books/030908251X/html/213.html

Pennsylvania Department of Environmental Protection—Glossary of mining terms
www.dep.state.pa.us/dep/deputate/minres/dms/website/training/glossary.html

United Nations—Glossary of environmental terms
http://www.nyo.unep.org/action/ap1.htm

U.S. Department of Energy/Energy Information Administration—Coal terms
http://www.eia.doe.gov/cneaf/coal/page/gloss.html

U.S. Department of Energy/Energy Information Administration—Energy terms
http://www.eere.energy.gov/consumerinfo/energyglossary.html

U.S. Environmental Protection Agency—Glossary of climate change terms
http://yosemite.epa.gov/oar/globalwarming.nsf/content/Glossary.html

U.S. Geological Survey—National water quality assessment glossary
http://water.usgs.gov/nawqa/glos.html

Virginia Division of Mineral Resources—Glossary of coal terms
www.mme.state.va.us/Dmr/DOCS/minres/coal/glos.html

Wyoming Coal Glossary
http://nasc.uwyo.edu/coal/library/lookup.asp

Credits

Front Cover — Cincinnati skyline (Corbis,); Blue Marble Earth (NASA); Coal (Digital Vision).

Inside front cover and Title page — Blue Marble Earth (NASA); Coal (Digital Vision).

Foreword and Preface — Mazonia-Braidwood Fish and Wildlife area in central Illinois (S. Greb, Kentucky Geological Survey).

Page 6 — Fig. 1, Energy sources: Power lines (Arch Coal, Inc); Charts (U.S. Data-Energy Information Administration, U.S. Department of Energy, 2003, World data from 2002); Pie charts- (S. Greb, C. Rulo, Kentucky Geological Survey).

Page 7 — Coal (Digital Vision); Fig. 2, World Coal Reserves: Earth (Digital Vision); Chart (Data-Energy Information Administration, U.S. Department of Energy, 2004, Chart- S. Greb, Kentucky Geological Survey).

Page 8 — Fig. 3, Coal, atomic structure: Coal (R. Busch); Molecule (S. Greb and C. Eble, Kentucky Geological Survey).

Page 9 — Fig. 4, Coal formation: Fern fossil, Painting of swamp with arthropods, Alaska peat bog, Florida sunset swamp (S. Greb, Kentucky Geological Survey). Fig. 5, Carbon cycle: (S. Greb, Kentucky Geological Survey).

Page 10 — Fig. 6, Greenhouse gases (Data- U.S. Environmental Protection Agency Greenhouse Gas Inventory, 2004); Fig. 7, Peat to Coal (S. Greb, Kentucky Geological Survey).

Page 12-13 — Fig. 8, Coal fields Map (Enhanced version of USGS map from GIS database); Coal Production Map & Coal Reserves Map (Data- Energy Information Administration, U.S. Department of Energy, State electricity profiles, 2003).

Page 14 — Fig. 9, Surface mine, Powder River basin, *large photo* (Peabody Energy, St. Louis, MO); Below ground- Twenty mile longwall mine, Colorado (Peabody Energy); Continuous miner cutting head (J. Ferm Collection, Kentucky Geological Survey). Thumbnails- Soil measurement in Texas (C.Meyers, Office of Surface Mining); Mine planning (S. Greb, Kentucky Geological Survey); Blasting, West Virginia (M. Blake, West Virginia Geological and Economic Survey); Reclamation of prime farm land, Indiana (C. Meyers, Office of Surface Mining).

Page 15 — Fig. 10, Drill rig (J. Cobb, Kentucky Geological Survey); Cores from a West Virginia mine (M. Blake, West Virginia Geological and Economic Survey).

Page 16 — Fig. 11, Mining Methods (Enhanced version of diagram by S. Greb, Kentucky Geological Survey).

Page 17 — Underground mining, Continuous miner cutting heads (J.Ferm Collection, Kentucky Geological Survey); Surface mining, Dragline shoveling at dusk (J. C. Cobb, Kentucky Geological Survey); Fig. 12, Large surface mine Powder River Basin, Wyoming (R. M. Flores, U.S. Geological Survey).

Page 18 — Fig. 13, Mountain-top removal mining in West Virginia (M. Blake, West Virginia Geological and Economic Survey).

Page 19 — Fig. 14, Sediment pond and mine in Kentucky (B. Davidson, Kentucky Geological Survey); Before and After in Virginia (Virginia Department of Mines, Minerals and Energy from the Office of Surface Mining); Recontouring a highwall in eastern Kentucky (C. Eble, Kentucky Geological Survey); Planting along Porcupine Creek at North Antelope Rochelle Mine, Wyoming (Peabody Energy); Returning mine land to grazing, Kentucky (C. Meyers, Office of Surface Mining).

Page 21 — Fig. 15, Restoration of native pine on mine lands, Black Mesa, Arizona (Peabody Energy); Fig. 16, Subsidence damage to a church, western Kentucky (J. Kiefer, Kentucky Geological Survey).

Page 22 — Fig. 17, Landslide at pre-1977 abandoned mine, eastern Kentucky, *photo at top* (J. Kiefer, Kentucky Geological Survey); Diagram (S. Greb, Kentucky Geological Survey); Returning highwall to original contour and building a retaining wall to prevent landslide damage (C. Meyers, Office of Surface Mining).

Page 23 — Fig. 18, Rock drain into sediment pond, Indiana; Ponds built during reclamation to provide flood storage in Texas (C. Meyers, Office of Surface Mining).

Page 24 — Fig. 19, Scientists counting fish in a stream on abandoned mine land (Kentucky Dept. of Natural Resources, Division of Abandoned Mine Lands); Water testing for pH in cup (S. Greb, Kentucky Geological Survey); Water background (Digital Vision); Recontouring arid lands after mining in New Mexico (State of New Mexico, Energy, Minerals and Natural Resources Department, Mining and Minerals Division).

Page 25 — Fig. 20, Acidic drainage, acidic water background (S. Greb, Kentucky Geological Survey); Sulfide mineral oxidation cycle (DeAtley Design, photos, USGS).

Page 27 — Fig. 21, Acidic drainage remediation, *large photo* (J. Skousen, West Virginia University).Thumbnails- Analyzing drainage (J. Skousen, West Virginia University); Directing acidic drainage in rock-lined ditch (Office of Surface Mining press release); Constructed wetland, Iowa (C. Meyers, Office of Surface Mining); Constructing a limestone drain (J. Skousen, West Virginia University); Vertical flow ponds (Office of Surface Mining press release). Fig. 22, Groundwater monitoring (D. Cumby, Kentucky Geological Survey).

Page 28 — Fig. 23, Coal mine fires, Smoke from underground coal fire coming through road, Centralia, Pennsylvania (M. Nolter); Warning sign, Centralia, Pennsylvania (J. Hower, Center for Applied Energy Research, University of Kentucky); Coal fire diagram (S. Greb, Kentucky Geological Survey, modified from U.S. Bureau of Mines).

Page 30 — Fig. 24, Collecting core for coalbed methane analyses (C. Eble, Kentucky Geological Survey); Methane pie chart (Data- U.S. Environmental Protection Agency Emissions Inventory, chart-S. Greb and C. Rulo, Kentucky Geological Survey).

Page 31 — Fig. 25, Public safety hazards, Kentucky surface mine next to house (C. Eble, Kentucky Geological Survey); Abandoned building at mine, Alaska (State of Alaska Dept. of Natural Resources Abandoned Mine Land Program); Flooding mine waters from blowout along Daniel Boone Parkway Kentucky (Kentucky Dept. of Natural Resources, Division of Mine Reclamation and Enforcement); Diagram (S. Greb, Kentucky Geological Survey); Dangerous highwall at abandoned mine, Alaska (State of Alaska Dept. of Natural Resources, Abandoned Mine Land Program); Mine safety sticker (Mine Safety and Health Administration).

Page 33 — Fig. 26, Mine safety, Surface mine, eastern Kentucky (B. Davidson, Kentucky Geological Survey); Continuous miner cutting head (J.Ferm Collection, Kentucky Geological Survey); Robot (Mine Safety and Health Administration press release); Fatalities chart (Data-Mine Safety and Health Administration, chart-modified from S. Greb, Kentucky Geological Survey, from a graph in Goode, 2002, *The Cutting Edge*).

Page 34-35 — Fig. 27, Large aerial photo of train in Wyoming (Peabody Energy). Thumbnails- Train (Digital Vision); Loading barge, Loading conveyors (S. Greb, Kentucky Geological Survey); Truck at Wyoming surface mine (Peabody Energy). Silo and train- Jacobs Ranch Mine in Wyoming's Powder River Basin (Kennecott Energy).

Page 36 — Fig. 28, Preparation plant (C. Meyers, Office of Surface Mining); Microscopic pyrite (C. Eble, Kentucky Geological Survey); Diagram (S. Greb, Kentucky Geological Survey); Magnetite recovery equipment, Heavy medium tanks (T. Miller, East Fairfield Coal Co., Ohio).

Page 37 — Fig. 29, Slurry discharge into impoundment in West Virginia (M. Blake, West Virginia Geological and Economic Survey).

Page 38 — Fig. 30, Pleasant View mine (Ketchup Lake) reclamation project, Kentucky, sequence of photos (Kentucky Dept. of Natural Resources, Division of Abandoned Mine Lands).

Page 39 — Fig. 31, Martin County, Kentucky, slurry images (Kentucky Dept. of Natural Resources).

Page 40 — Fig. 32, Dredging slurry to recover coal in western Kentucky (S. Greb, Kentucky Geological Survey). Fig. 33, Slurry pond reclamation sequence of photos (Kentucky Dept. of Natural Resources, Division of Abandoned Mine Lands).

Page 41 — Fig. 34, Abandoned, pre-1977 gob pile, central Illinois (S. Greb, Kentucky Geological Survey).

Page 42 — Fig. 35, Large photo of power plant (Corbis); Coal Usage pie chart (Data- Energy Information Administration, 2004 Annual Report); Electricity from coal map (Data- Energy Information Administration, 2003 State electricity profiles, map-S.Greb, Kentucky Geological Survey).

Page 43 — Fig. 36, Power plant diagram (Modified by S. Greb from Tennessee Valley Authority).

Page 45 — Fig. 37, Acid rain diagram (Modified by S. Greb from U.S. Environmental Protection Agency and other sources). Fig. 38, Sulfate maps (U.S. Environmental Protection Agency: Trends in Wet Sulfate Deposition Following Implementation of Phase I of the Acid Rain Program).

Page 46 — Fig. 39, SO_2 graph (SO_2 data-U.S. Environmental Protection Agency, graph-S. Greb, Kentucky Geological Survey). Fig. 40, Scrubber diagram (Modified by S. Greb from U.S. Environmental Protection Agency).

Page 47 — Fig. 41, NO_x burner diagrams (Modified by S. Greb from U.S. Dept. of Energy).

Page 48 — Fig. 42, Baghouse and ESP diagrams (Modified by S. Greb from U.S. Environmental Protection Agency diagrams). Fig. 42, Particle matter size diagram (U.S. Dept. of Energy). Fig. 44, Mercury bioaccumulation diagram (S. Greb, Kentucky Geological Survey).

Page 49 — Fig. 45, Mercury sources pie chart (Redrafted from U.S. Dept. of Energy). Fig. 46, Greenhouse effect diagram (Modified from Environmental Protection Agency diagram by S. Greb).

Page 50 — Fig. 47, CO_2 Pie chart (Data-U.S. Dept. of Energy, chart-S. Greb). Fig. 48, CO_2 Injection well diagram (S. Greb, Kentucky Geological Survey).

Page 51 — Fig. 49, Wallboard and diagram of gypsum byproduct production (S. Greb, Kentucky Geological Survey).

Page 52-53 — Fig. 50, Miner at Spring Creek Mine in Montana (Kennecott Energy). Mazonia-Braidwood Fish and Wildlife area in central Illinois (S. Greb, Kentucky Geological Survey).

Page 54 — Fig. 52, Mountains (Digital Vision); Coal texture background (Corbis).

Page 55 — Power lines (Corbis). Fig. 53, Fluidized Bed Combustor unit diagram (Modified from U.S. Dept. of Energy diagram by S.Greb).

Page 56 — Fig. 54, Integrated Gasification Combined Cycle diagram (Modified from U.S. Dept. of Energy by S. Greb).

Page 59 — Blue Marble Earth (NASA); Coal (Digital Vision).

Inside back cover — Background photo (Digital Vision).

Back cover — Wetland, Iowa (C. Meyers, Office of Surface Mining); Coal (Digital Vision).

Index

a abandoned mine lands, 22, 27-28, 31-32
acidic drainage, 25-27, 37-38
acid rain, 44-47
auger mining, 17

b black lung disease, 32-33
blasting, 30-32
blowouts, 31-32

c carbon, 8-10
carbon cycle, 8-9
carbon dioxide, 9-10, 30, 49-51
clean air legislation, 45-47, 49, 54
clean-coal technology, 46, 55
Clean Water Act, 21, 54
cleats, 11
coalfields, 12-13
coal-fired power plants, 42-44
coal formation, 7-11
coal mine fires, 18, 28-29
coal swamp, 8
contour mines, 17
coring, 15, 30

d dewatering, 28, 37
distribution, 12-13
drilling, 15
dust control, 30-33

e electrostatic precipitators, 48
emissions, 8, 10, 28-29, 44-49
energy sources, 6
environmental concerns/impacts, 7-8, 16, 18-33, 37-41
environmental protection, 54
erosion, 23-24
exploration, 15-16

f fatalities, 32-33
filter baghouse, 48
flue-gas desulfurization, 46
fluidized bed combustion, 41, 46-47, 55-56
flooding, 18-19, 23-24
fossil fuel, 7-9
FT synfuels, 58
fugitive methane, 18, 29-30
FutureGen, 57-58

g gasification, 43, 47, 50-51, 56-58
geologic basins, 10-13
gob, 37
gob fires, 37, 41
greenhouse gases, 9-10, 29, 46-47, 49-51
groundwater protection, 27-28, 39-40

h hazardous air pollutants, 48-49
health and safety, 18, 30-33, 47-49
highwall mining, 17-23, 31

i impoundments, 23, 37-41
impurities, 25-26, 34-35
integrated gasification combined cycle, 56-57

k Ketchup Lake remediation, 38-39

l landslides, 18-19, 22-23
limestone drains, 25-27
liming, 38-39
longwall mining, 22

m mercury, 44, 48-49
methane, 10, 29-30
mine safety, 18, 32-33
mires, 10
mining cycle, 14-15
mining methods, 16
mountaintop removal, 18, 20-21

n nitrogen oxides, 8, 44-47

p particulate emissions, 47-48
peat, 8-11,
permits, 15-16, 20
physical disturbance, 18-21, 37, 41
post mine land use, 19
power and heat generation, 7, 42-44
processing, 34-41
production, 7, 11-12, 17
public safety, 18, 30-32, 37
pyrite, 25-26, 36-37

r rail transport, 34-35
rank, 10-11
reclamation, 19-21, 23, 37, 40-41
regrading, 19, 38
regulations, 21-23, 30, 54
remediation, 26, 38-39
resources and reserves, 7, 11-13, 52-53, 58-59
revegetation, 19-21, 38
road damage, 37-38
room- and pillar-mining, 22
runoff, 23-24, 39-40

s scrubbers, 46-47, 53, 55
sediment ponds, 23
selective catalytic reduction, 47
sequestration, 50-51, 59
silicosis, 32-33
slurry ponds, 37, 39-41
solid waste byproducts, 46, 51
spoil, 24, 26
subsidence and settlement, 18, 21-22, 28-29
sulfur, 8
sulfur dioxide, 8, 44-46
surface mining, 17-21
Surface Mining Control and Reclamation Act, 21, 54
surface water protection, 20-21, 24-27
syngas, 43

t transportation, 34-38

u underground mining, 16-17, 20
usage, 7, 42-51

w water quality and protection, 18, 20-21, 24-28, 37-40